Proceedings

44 International Congress of Americanists
Congreso Internacional de Americanistas
Manchester 1982

General Editor: Norman Hammond

# Drained Field Agriculture in Central and South America

edited by

J. P. Darch

BAR International Series 189
1983

# B.A.R.

122 Banbury Road, Oxford OX2 7BP, England



B.A.R.S189, 1983 :'Drained Field Agriculture in Central and South America'



Price £ 16.00 post free throughout the world. Payments made in currency other than sterling must be calculated at the current rate of exchange. Cheques should be made payable to B.A.R. and sent to the above address.

ISBN 0 86054 241 6

For details of all BAR publications in print please write to the above address. Information on new titles is sent regularly on request, with no obligation to purchase.

Volumes are distributed direct from the publisher. All BAR prices are inclusive of postage by surface mail anywhere in the world.

Printed in Great Britain

Frontispiece: Location of drained fields in Central and South America

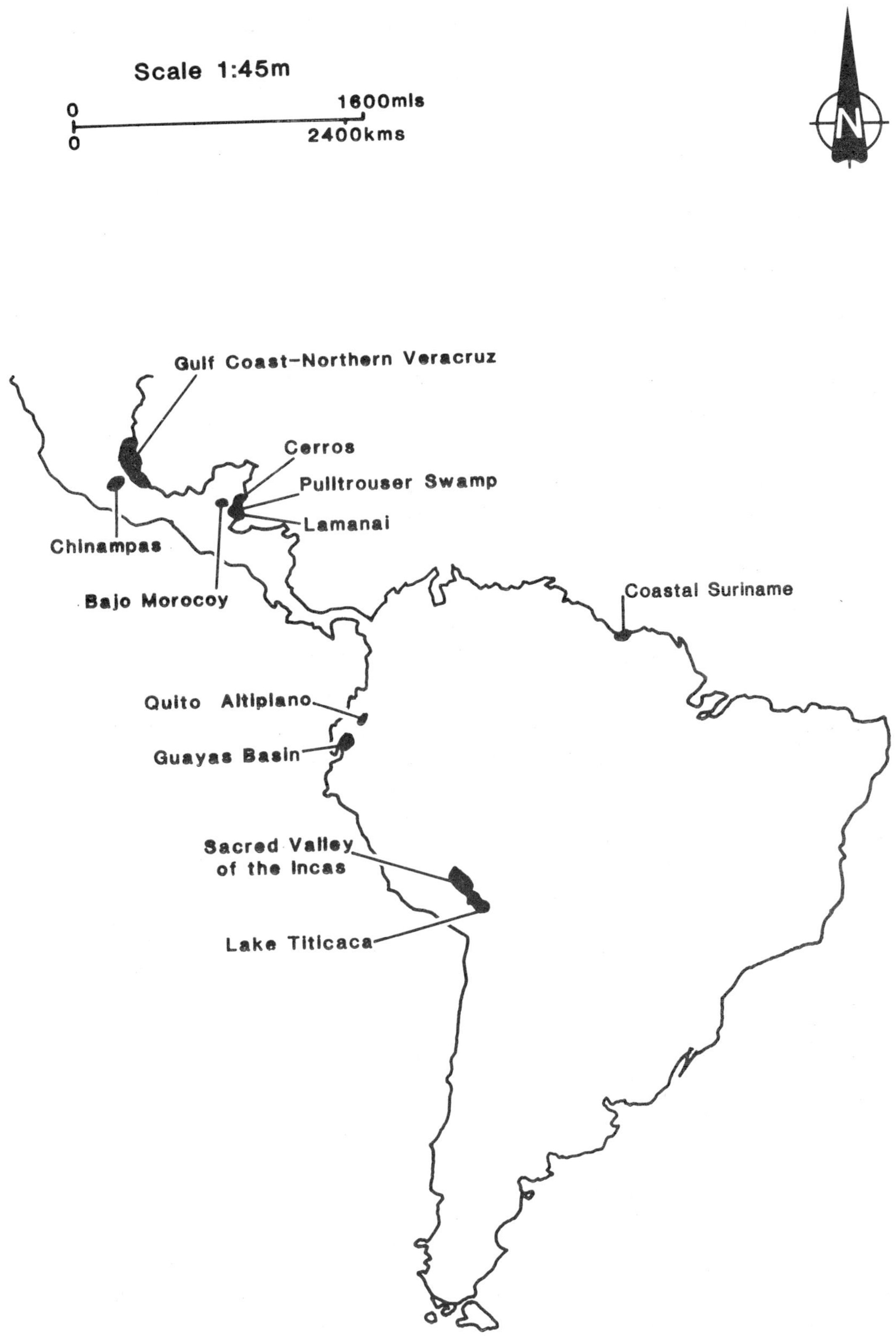

## CONTENTS

## LIST OF FIGURES

## LIST OF TABLES

## LIST OF CONTRIBUTORS

M.F. Amador, Colegio Superior de Agricultura Tropical, Cardenas, Tabasco, Mexico.

J.T. Arnason, Department of Biology, University of Ottawa, Ottawa, Canada, K1N 6N5.

J.P. Darch, School of Environmental Sciences, University of East Anglia, Norwich, NR4 7TJ, England.

W.M. Denevan, Department of Geography, University of Wisconsin, Madison, Wisconsin 53705, U.S.A.

I.S. Farrington, Department of Prehistory and Anthropology, Australian National University, Canberra, A.C.T., Australia, 2600.

S.R. Gliessman, Environmental Sciences, University of California, Santa Cruz, California, 95064, U.S.A.

G. Knapp, Department of Geography, University of Minnesota, Minneapolis, 55455, U.S.A.

J.D.H. Lambert, Department of Biology, Carleton University, Ottawa, Canada, K1S 5B6.

T.J. Lennon, Western Cultural Management Inc., P.O. Box 2326, Boulder, Colorado, 80306, U.S.A.

K. Mathewson, Department of History and Geography, Virginia Commonwealth University, Richmond, Virginia 23284, U.S.A.

F.J. Rosado May, Colegio Superior de Agricultura Tropical, Cardenas, Tabasco, Mexico.

T. Rojas Rabiela, Centro de Investigaciones y Estudios Superiores en Antropologia Social, Hidadgo y Matamores, Apdo Postal 22-048, Tlalpan, Mexico.

R. Ryder, Centro Panamericano de Estudios e Investigaciones Geograficas, Quito, Ecuador.

V. Scarborough, Department of Anthropology, University of Texas - El Paso, El Paso, Texas, 79963, U.S.A.

A.H. Siemens, Department of Geography, University of British Columbia, Vancouver, British Columbia, Canada, V6T 1W5.

R.T. Smith, School of Geography, University of Leeds, Leeds, LS2 9JR, England.

B.L. Turner II, Graduate School of Geography, Clark University, Worcester, Massachusetts, 01610, U.S.A.

A.H. Versteeg, Instituut Voor Prehistorie der Rijksuniversiteit te Leiden, Leiden, The Netherlands.

S. Jeffrey K. Wilkerson, The Institute for Cultural Ecology of the Tropics, Inc., P.O. Box 3324, Tampa, Florida, 33601, U.S.A.

## Preface

The origins of this volume were in the Drained Fields of the Americas Symposium held at the 44th International Congress of Americanists in Manchester, England, in September 1982. The half day symposium included the papers by Versteeg, Smith, Siemens, Darch, Knapp and Ryder and Lambert and Arnason while the remainder of the collection were invited contributions for the book. The fourteen papers are not a comprehensive survey of all the known drained field systems in Central and South America, rather, they are reports and assessments of a selection of ongoing and recent research projects. Therefore the volume has a certain bias towards the Central American countries of Belize and southern Mexico where vast areas of lowland drained fields have recently been reported. The South American field systems discussed are detailed studies of both highland and lowland fields that have previously been discovered, although the extent of the systems is often greater than was originally documented.

Controversy surrounds the naming and definition of Pre-Hispanic intensive agricultural systems. For this book the term "drained field" has been favoured because the name does not imply a particular field form, as raised field or ridged field does. It should be recognised though that some types of fields discussed in this volume are not drained *per se*, for instance, they may have been constructed primarily for climatic manipulation. However, "drained fields" describes a majority of the fields discussed, but individual authors have been given licence to define their own terms if they so wished.

The production of the book would not have been possible without the help of many people. I would like to thank all of the contributors for their co-operation and readiness to deal with my editorial comments even from far flung parts of the world. I am also indebted to the B.A.R. Series editor of the 44th I.C.A. volumes and to B.A.R. for their advice and help in the preparation of the manuscripts. Barbara Slade courageously undertook the typing and Lucila Recart Haynes the Spanish translations, to both I am very grateful. Finally I would like to express my gratitude to Richard Smith for his support in organising the symposium and encouragement in the production of the book.

Norwich 1983 J.P. Darch

# Prólogo

Este tomo se originó en el Simposio sobre Campos Drenados de las Américas que se efectuó en el 44º Congreso Internacional de Americanistas en Manchester, en septiembre de 1982. El simposio de medio día incluyó los trabajos de Versteeg, Smith, Siemens, Darch, de Knapp y Ryder, y Lambert y Arnason, mientras que el resto de la colección fueron contribuciones invitadas especialmente para el libro. Los catorce trabajos no constituyen un estudio global de todos los sistemas de campos drenados que se conocen en Centro y Sud América. Más bien son informes y contribuciones sobre una selección de proyectos de investigación actuales y recientes. Por lo tanto el tomo se inclina en cierta medida hacia los países centroamericanos de Belice y el sur de México donde recientemente se ha informado sobre areas extensas de campos drenados de tierras bajas. Los sistemas de campos sudamericanos que se tratan aquí son estudios detallados de campos tanto de tierras altas como bajas que han sido estudiados previamente, aunque la extensión del sistema es a menudo mayor de lo que se indicó originalmente.

Existe cierta controversia sobre el nombramiento y definición de los sistemas agrícolas intensivos prehispánicos. En este libro se ha preferido el término "campo drenado" porque el nombre no implica cierta forma especial de campo, tal como lo hacen campo elevado o campo alomado. Pero debe reconocerse que algunos tipos de campos tratados en este tomo no son drenados *per se*, por ejemplo, pueden haberse construído originalmente para el manejo climático. Sin embargo, "campos drenados" describe la mayoría de los campos tratados pero a cada autor se le ha permitido definir sus própios términos si así lo prefiere.

La producción del libro no habría sido posible sin la ayuda de muchas personas. Quisiera agradecerle a todos los contribuyentes por su cooperación y su buena disposición para atender a mis comentarios editoriales aun desde lugares remotos de la tierra. También le estoy agradecida al redactor de la serie B.A.R. (British Archaeological Reports) de los tomos del 44º I.C.A. (International Congress of Americanists) y a B.A.R. por sus consejos y ayuda en la preparación de los manuscritos. Barbara Slade se comprometió heroicamente a ejecutar la dactilografía y Lucila Recart Haynes a traducir al español; a ambas les agradezco. Finalmente quisiera agradecerle a Richard Smith por su apoyo en organizar el simposio y su aliento en la producción de este libro.

Chapter 1

# DRAINED FIELDS IN THE AMERICAS : AN INTRODUCTION

J.P. Darch

Abstract

Drained fields are defined as a form of intensive wetland agriculture on artificially constructed platforms located in areas of permanent or seasonal inundation. Pre-Hispanic drained fields built by the Maya, Inca, Guaya and Aztecs among others are discussed. The construction of contemporary drained fields to increase the food resources of present-day village communities are also cited. The New World drained fields temporally span several hundred years, the earliest possibly dating from ca. 2000 B.C., while some highland Mexican fields in use at the Conquest are still cultivated today. Spatially the distribution and area of drained fields is not fully documented and over the past decade or so extensive areas of fields have been located and added to the total area mapped. It seems that drained fields were an important form of New World agriculture that could yield enough food for the high population densities estimated for such areas as the Maya Lowlands. Any doubts that the drained fields are in fact natural soil features has now been discounted.

Resumen

Los campos drenados se definen como una forma intensiva de agricultura de tierras húmedas sobre plataformas construídas artificialmente ubicadas en areas de inundación permanente o estacional. Se comentan los campos drenados prehispánicos construídos por los mayas, los incas, los guayas y los aztecas entre otros, y se citan ejemplos de la construcción contemporánea de campos drenados para aumentar los recursos alimenticios de las comunidades aldeanas actuales. Los campos drenados del Nuevo Mundo abarcan varios cientos de años, los más antiguos remontándose hasta alrededor de 2000 A.C., mientras que algunos campos mexicanos de tierras altas que estaban en uso durante la Conquista aun se cultivan hoy día. La distribución espacial y el area de los campos drenados no esta plenamente documentada y durante la última década extensas areas de campos se han localizado y agregado al area total cartografiada. Al parecer los campos drenados constituyeron una forma importante de la agricultura del Nuevo Mundo, capaz de producir suficiente alimento para las altas densidades de población calculadas para areas tales como las tierras bajas maya. Se ha desechado cualquier duda sobre si los campos drenados son, en efecto, rasgos naturales del suelo.

## Introduction

Prehispanic or contemporary forms of intensive wetland agriculture involving the creation of cultivation platforms surrounded by water on a permanent or seasonal basis, are collectively called drained fields. The features are also known as raised or ridged fields, channelized fields, tablones, camellones or chinampas, when the names denote or describe particular forms of drained field. Often the construction of the fields involved elevating parts of the surface of floodplains, swamps, lake margins or shallow lakes above water level. This entailed carting and tipping vast quantities of fill material to act as ballast. The fill often had to be transported some distance to the site, thus involving considerable physical effort. It can therefore be deduced from the structural remnants of drained fields that the labour involved in building them was immense (Turner, Chapter 2) and that the organisation of the labour and the attention given to details of the associated hydrological regimes demanded highly skilled and sophisticated central planning. Stemming from these deductions and from contemporary attempts to build drained fields (Gomez-Pompa 1982) it is logical to assume that the initial effort of construction warranted intensive, year-round cultivation of high-yielding crops. Among the crops which have been suggested as suitable for drained field cultivation are maize, beans, squash, cacao and cotton (Rojas Chapter 8). Pollen and macrofossil remains of some of these crops have been found, for instance in the Pulltrouser Swamp fields (Turner and Harrison 1981). On the various contemporary drained fields that have been built in Mexico such crops as manioc, cabbage, squash, rice, corn, water-melon, alfalfa, chile, carrots, turnip and salad vegetables have been successfully grown (Gomez-Pompa 1982). Pisciculture has flourished in the canals of the new Nacajuca field system near Villahermosa, Mexico, too, and it is likely that the ancient canals of Prehispanic drained fields were used in much the same way, thus adding a source of protein to supplement the crops grown on the fields.

Over the past decade or so, much data has been added to our knowledge of drained field agriculture in the New World. Since the first reports of drained fields in the Llanos de Mojas in north eastern Bolivia in the early 1900's (Nordenskiold 1913; 1916) and the later documentation by Plafker (1963) and Denevan (1963; 1966) many more reports of drained fields have been made. Over the past two decades extensive areas of drained fields have been identified - associated with the Maya and Aztecs in Central America, and the Incas and the Guayas in South America. Other fields have been found in the Central Orinoco Llanos, Venezuela, coastal Suriname, the San Jorge region of northern Colombia, the Calima Valley of Colombia, Argentina and Brazil (Denevan 1970; 1982). Some of these field systems are discussed in this volume where the aim has been to draw together recent parallel research on drained field agriculture from various parts of the continent.

## Distribution of Drained Fields

Until comparatively recently drained field systems were thought to occur mainly in savanna grassland regions such as those reported from Beni State, Bolivia (Denevan 1960), but most reports of relic fields originally came from such areas because spotting field outlines by ground or aerial survey is easier in grassland than in forested areas. However, predictions that drained fields also occurred under forest were made for some areas, in particular the Maya lowlands of Central America (Turner 1974; Harrison 1975). In many places, particularly in the densely forested, internally-drained alkaches or *bajos* of the Maya area, grid-like patterns in the vegetation, suggestive of drained fields masked by forest, could be seen from the air when sun angles were low (Turner and Harrison 1978). Rectangular forested structures were also noted ringing *bajos* such as Pulltrouser Swamp, Belize (Siemens 1977). However, until 1979 none of the forested drained fields surrounding swamps had been excavated, although floodplain drained fields had been confirmed on the Río Candelaria, Campeche, Mexico, by Siemens and Puleston (1972) and at Albion Island on the Río Hondo, Belize (Siemens 1977). The Pulltrouser Swamp project, directed by Turner and Harrison (Turner and Harrison 1981; 1983), confirmed that the patterns seen from the air around this *bajo* were indeed Maya drained fields and not natural soil patterning caused by gilgai formation in clay soils prone to extremes of shrinking and swelling. The fields contained confirmative evidence in the form of Maya ceramics, lithic tools, crop pollen and macrofossils and dateable wood and charcoal. Subsequently, ground observations and excavation of vegetation patterns at the Bajo Morocoy, at Nicolas Bravo, Quintana Roo, Mexico, also revealed a drained field system (Gliessman *et al*. Chapter 5) which Turner had predicted (Turner 1974).

Concomitant with this confirmatory archaeological and ecological evidence of drained field systems around forested swamps, new remote sensing techniques using various forms of radar imagery began to reveal many more new sites of drained fields in lowland Guatemala, Belize and Mexico (Adams *et al*. 1981). Some initial data came from the processing of Seasat imagery but more convincing results at a scale of 1:250,000 resulted from SLAR (side-looking airborne radar) imagery flown over the area. Scarborough (Chapter 7) reports the location of canals at the northern Belize site of Cerros from the same imagery. Other parts of the New World where extensive areas of drained fields have recently been located are the Guayas Basin of Ecuador (Denevan and Mathewson Chapter 9), the coastal plain of western Suriname (Versteeg Chapter 13), the Quito Altiplano (Knapp and Ryder Chapter 11), Lake Titicaca, Peru (Lennon Chapter 10) and coastal Veracruz, Mexico (Wilkerson Chapter 4). Therefore the drained field systems recently reported are from both lowland and highland regions, coastal and inland sites, and river and lacustrine locations, thus suggesting that this agricultural form could be adapted for use in many different environmental situations.

The temporal distribution of drained fields is more difficult to assess than their spatial distribution because the archaeologist has to depend on dateable remains found in or associated with the field system to rank them. Intensively used drained fields tend not to contain large amounts of ceramic debris and that which exists has often been stratigraphically mixed by cultivation. Dateable organic material is often rare too because constant cultivation and fluctuating soil water levels have introduced air into the soils and hence promoted the oxidisation of organic debris (Smith Chapter 14). However, some field systems have been dated. It seems that the earliest fields so far recorded are in the Guayas Basin of Ecuador, where soils associated with drained fields date from ca. 2000 B.C. (Denevan and Mathewson Chapter 9). Maya fields are later. Turner (Chapter 2) suggests that some of the Maya fields at Pulltrouser may be as early as 1000 B.C. but most date from A.D. 300-1000, as do those at Lamanai, Belize, discussed by Lambert and Arnason (Chapter 6). The fields on the coast of Veracruz, Mexico, investigated by Wilkerson (Chapter 4), may date between A.D. 1 and 600, while those of the Quito Altiplano, studied by Knapp and Ryder (Chapter 11), are probably associated with the Chillogallo settlement which was occupied between A.D. 1250-1525. Likewise, the fields in coastal Suriname have only been dated by association with nearby habitation mounds, which indicates that they were probably in use between ca. A.D. 150-700 (Versteeg Chapter 13). Therefore, since the early 1970's it has emerged that drained field agriculture in the New World tropics was an areally extensive and important agricultural technique temporally spaced over several hundred years, and may be originating in the Guayas Basin of Ecuador.

## Natural Features versus Drained Fields

It seems that the recent excavations of the vegetation patterns at both Pulltrouser Swamp and Bajo Morocoy in the Maya Lowlands have laid to rest, in this area at least, the view that the ground patterns are natural features. Two possible natural origins of the grid-like patterns in the vegetation were posited. The first, gilgai, has already been mentioned, the notion being proposed by Puleston (1978) among others. The second, that of joints in the limestone bedrock influencing vegetation stature by causing differential soil moisture availability was also discussed (Siemens 1978; Scarborough Chapter 7). The gilgai debate arose because rectangular drained fields, surrounded by ditches, look remarkably like tank-shaped gilgai as described from Australia by Prescott (1931) and Hallsworth *et al.* (1955). Also, the seasonally wet and dry climatic conditions characteristic of many drained field sites are ideal for gilgai formation. However, detailed soil investigations at Pulltrouser Swamp, Belize, in 1979, showed that although the soil possesses some of the properties that could contribute to gilgai formation (high montmorillonite clay content, high sodium levels and slightly alkaline pH values) the present dry season was insufficiently intense to produce gilgai (Darch 1979; 1983). Bouyoucos Blocks installed in the soils of the drained fields during the 1979 dry season indicated that the clays remained in an expanded state throughout the dry season at depths greater than 10 cm. Extrapolating these results south and west of Pulltrouser into areas where ground patterns also occur, as at Tikal, Guatemala, (Olson 1969) we find that in these areas the mean annual rainfall is higher than at Pulltrouser and the dry season is less intense and shorter. Combined, these factors dismiss the likelihood of the patterns being gilgai.

Less work has been done to investigate the influence of bedrock geology on vegetation patterning. Such patterning could produce a regular grid pattern which might influence soil drainage and vegetation growth and hence be detected by aerial observations, including the imaging radar recently flown over the Maya Lowlands. To date, there exists insufficient evidence to prove or disprove the role of bedrock geology. Confirmation or denial of drained fields at the sites located by the radar imagery will have to await the results of the field checking programme.

## Drained Field Classification

In this volume the term drained field is used to embrace a range of intensive agricultural systems adapted to a variety of aquatic environments. However, it is probably wise to view drained fields as representing the aquatic end of a spectrum of intensive agricultural techniques which extend from the sunken fields and canalized rivers in areas of low moisture availability, as on the Peruvian coast, (Parsons and Psuty 1975; Smith 1979; Farrington Chapter 12), to chinampas built in shallow lakes as in the Valley of Mexico (Armillas 1971). In terms of drained fields *per se* it is suggested that the forms span channelized fields, where drainage ditches were excavated into swamps and possibly floodplain margins, through raised fields where cultivation platforms were elevated above seasonal inundation levels, to chinampas built in shallow lakes. In each case the aim was to produce a well drained, aerated, well structured and fertile soil capable of producing high crop yields over long time periods given good management techniques. The various types of drained field are thus adaptations to a variety of hydrological regimes ranging from seasonally arid to permanently inundated.

## Shape and Pattern of Drained Fields

The organisation of individual drained field systems can reveal a great deal about the hydrological regime within which each set of fields was built to function. Of particular interest is recent work by Lennon (1982; Chapter 10) on the Peruvian drained fields surrounding Lake Titicaca. Here a statistically proven relationship has been found between field size characteristics and distance to either the associated lake or river system; riverine fields are larger than lacustrine fields. In particular, distance to hydraulic regime was seen to exert an influence on field patterns. In addition, the lake and riverine field patterns were distinct from each other. The data suggests that for both lake and riverine fields, patterning indicates an understanding of a range of natural hazards, notably flooding and drought. In addition, Siemens (1983; Chapter 3) notes preferred orientations of the long axis of drained fields in Central America, orientations which appear to relate to ceremonial centres as distinct from agricultural site considerations.

Drained field shapes and sizes vary enormously (Denevan 1970; 1982) according to local conditions. Coupled with this, each field must function as a separate hydrological entity to some extent, each field developing a water table level in accordance with its particular dimensions and soil texture (Smith Chapter 14). It therefore seems increasingly clear that the characteristics of individual drained field systems indicate an element of adaptation to local hydrological and soil moisture regimes.

## The Role of Drained Fields in Agriculture

It can be hypothesised with some confidence that drained field construction reflected the need to intensify agricultural production probably in the face of population pressure when other land use strategies were no longer capable of sustaining a rising population (Turner 1978; Chapter 2). The apparent success of drained field agriculture in Prehispanic times has encouraged research into the construction and development of contemporary drained fields, particularly in Mexico (Gomez-Pompa 1978; Gomez-Pompa *et al*. 1982; Gliessman 1979; Gliessman *et al*. 1981. Experimentation began in the state of Tabasco in the agriculturally devastated Balacan-Tenosique area. Chinampas were built on an experimental basis in 1976. From this experiment the idea of contemporary chinampa agriculture spread. In Mexico alone further fields were built on lowland swampy sites at El Espino, Cardenas, Nacajula and El Morrode de Manche, Tabasco, and at a highland site, Tecocomulco, in Hidalgo State. The benefits of chinampa cultivation are that they are constructed and gardened by local villagers to produce food for local consumption in areas marginal for agriculture; this contrasts with many agricultural improvement schemes where large landowners are brought in to mechanically cultivate commercial crops for export. Cultivation is intense so enough food can be grown to sustain the local population and because canal muck is used as manure the expense of fertilizers is spared.

An understanding of soil processes in relation to drained field agriculture is of considerable importance not only in explaining why these systems remained viable for substantial periods in the past but also as a basis for redeveloping them in the present century under appropriate economic circumstances. It seems that the builders of the field systems investigated were well grounded in hydrological manipulation (Lennon Chapter 10) and also in soil chemistry (Smith Chapter 14). Nevertheless, there are likely to have been large areas inconducive to such developments, where attempts to develop drained fields would have met with failure. Alternate forms of agriculture may have been practiced in such areas, possibly including slope terracing and slash and burn (see Turner Chapter 2).

## Conclusion

Drained fields in the New World are proving to be far more extensive than they were once thought to be. In the light of new exploration techniques the status of drained field agriculture must be regarded as an areally extensive, sophisticated form of adaptation by high population pressures to complex hydrological conditions. Drained fields are clear evidence of the control of wet season inundation and dry season droughts which when combined with activities of good management have produced a very lucrative form of agriculture, certainly an answer to the food requirements of large populations. Although studies of drained field distribution and ground plan tell us much about the water regime and the implied success of the systems, no approach so far has explained why other areas of similar terrain show an absence of fields.

## Bibliography

ADAMS, R.E.W., W.E. BROWN,Jr. and T. PATRICK CULBERT 1981, Radar mapping, archaeology and ancient Maya land use, Science 213, 1457-1463.

ARMILLAS, P. 1971, Gardens on swamps, Science 174, 653-661.

DARCH, J.P. 1979, Soils and Geomorphological Studies in Northern Belize, Central America, M.A. thesis, University of Oklahoma, 182pp.

DARCH,J.P. 1983, The soil, In, Pulltrouser Swamp: Ancient Maya Habitats, Agriculture and Settlement in Northern Belize, edited by B.L. Turner II and P.D. Harrison, University of Texas Press, Austin, Texas.

DENEVAN, W.M. 1963, Additional comments on the earthworks of Mojos in northeastern Bolivia, American Antiquity 28, 540-545.

DENEVAN, W.M. 1966, The aboriginal cultural geography of the Llanos de Mojos of Bolivia, Ibero Americana 48, University of California, Berkeley.

DENEVAN, W.M. 1970, Aboriginal drained-field cultivation in the Americas, Science 169, 647-654.

DENEVAN, W.M. 1982, Hydraulic agriculture in the American tropics: forms, measures, and recent research, In, Maya Substance: Studies in Memory of Dennis E. Puleston, edited by Kent V. Flannery, Academic Press, London, 181-203.

GLIESSMAN, S.R. 1979, Some ecological relationships of traditional agroecosystems in the lowland humid tropics of Southern Mexico, paper presented at XLIII International Congress Americanists, Vancouver, B.C. Canada.

GLIESSMAN, S.R., R. ESPINOSA, E. and M. AMADOR, A. 1981, The ecological basis for the application of traditional technology in the management of tropical agroecosystems, Agro-ecosystems 7, 173-185.

GOMEZ-POMPA, A. 1978, An old answer to the future, Mazingira 5, Pergamon Press, Oxford.

GOMEZ-POMPA, A., H.C. MORALES, E.J. AVILLA and J.J. AVILLA 1982, Experiences in traditional hydraulic agriculture, In, Maya Subsistence: Studies in Memory of Dennis E. Puleston, edited by Kent. V. Flannery, Academic Press, London, 327-342.

HALLSWORTH, E.G., G.K. ROBERTSON and F.R. GIBBONS 1965, Studies in pedogenesis in New South Wales, VII: The gilgai soils, Journal Soil Science 6, 1-31.

HARRISON, P.D. 1975, Intensive agriculture in Southern Quintana Roo, Mexico: some new lines of evidence and implications for Maya prehistory, paper presented at the XL Annual Meeting of the Society of American Archaeology, Dallas, Texas.

LENNON, T.J. 1982, Raised Fields of Lake Titicaca, Peru: A Prehispanic Water Management System, Ph.D. dissertation, University of Colorado, 340pp.

NORDENSKIOLD, E. 1913, Urnengraber und mounds im bolivianischen Flacklande, Baessler-Archiv, Band 3, 205-255, Liepzig.

NORDENSKIOLD, E. 1916, Die Anpassung der Indianer an die verhaltnisse in den Uberschivemmungsgebieten in Sudamerika, Ymer 36, Haft 2, 138-155, Stockholm.

OLSON, G.W. 1969, Description and data on soils of Tikal, El Peten, Guatemala, Central America, Mimeo 69-2, Dept. of Agronomy, Cornell University.

PARSONS, J.R. and N.P. PSUTY 1975, Sunken fields and Prehispanic subsistence on the Peruvian coast, American Antiquity 40, 259-282.

PLAFKER, G. 1963, Observations on archaeological remains in northeastern Bolivia, American Antiquity 28, 372-378.

PRESCOTT, J.A. 1931, The Soils of Australia in Relation to Vegetation and Climate, Council of Scientific and Industrial Research in Australia; Bulletin, 52, Canberra.

PULESTON, D.E. 1978, Terracing, raised fields and tree cropping in the Maya Lowlands: a new perspective on the geography of power, In, Prehispanic Maya Agriculture, edited by Peter D. Harrison and B.L. Turner II, University of New Mexico Press, Albuquerque, New Mexico, 225-245.

SIEMENS, A.H. 1977, Some patterns seen from the air, Journal Belizean Affairs 5, Special Issue: The Rio Hondo Project, 5-21.

SIEMENS, A.H. 1978, Karst and the Prehispanic Maya in the southern lowlands, In, Prehispanic Maya Agriculture, edited by Peter D. Harrison and B.L. Turner II, University of New Mexico Press, Albuquerque, New Mexico, 117-143.

SIEMENS, A.H. and D.E. PULESTON 1972, Ridged fields and associated features in Southern Campeche: new perspectives on the lowland Maya, American Antiquity 37, 228-239.

SMITH, R.T. 1979, The development and role of sunken field agriculture on the Peruvian coast, Geographical Journal 145, 387-400.

TURNER, B.L. II 1974, Prehistoric intensive agriculture in the Maya lowlands, Science 185, 118-124.

TURNER, B.L. II 1978, Ancient agricultural land use in the Central Maya Lowlands, In, Prehispanic Maya Agriculture, edited by Peter D. Harrison and B.L. Turner II, University of New Mexico Press, Albuquerque, New Mexico, 163-183.

TURNER, B.L. II and P.D. HARRISON 1978, Implications from agriculture for Maya Prehistory, In Prehispanic Maya Agriculture, edited by Peter D. Harrison and B.L. Turner II, University of New Mexico Press, Albuquerque, New Mexico, 337-373.

TURNER, B.L. II and P.D. HARRISON 1981, Prehispanic raised-field agriculture in the Maya lowlands, *Science* *213*, 399-405.

TURNER, B.L. II and P.D. HARRISON (editors) (1983), *Pulltrouser Swamp: Ancient Maya Habitat, Agriculture and Settlement in Northern Belize*, University of Texas Press, Austin, Texas.

Chapter 2

# CONSTRUCTIONAL INPUTS FOR MAJOR AGROSYSTEMS OF THE ANCIENT MAYA

B.L. Turner II

## Abstract

The problems of estimating crop yields from abandoned aboriginal agrosystems are discussed. Interpretation of those systems can be based on constructional inputs. Constructional inputs of drained fields are compared with hill terracing and dry-land fallowing in the Maya Lowlands. Drained fields are the most costly endeavour and they must have been associated with a high demand for production and a land intensive economy. Construction data for drained fields are taken from Pulltrouser Swamp, Belize, where it is estimated that construction took between 833 and 3116 work days per hectare. Terracing was another Classic Period agrotechnology in the Maya area. Terrace construction can be calculated in terms of linear metres or cubic metres, but a major factor in controlling building rates is the distance building stones have to be hauled. Cubic metre rates of $1.4m^3$/work day or 267 work days/ha and linear metre rates of 105 to 848 work days/ha are estimated for constructing broad base terraces in the Rio Bec area of southern Quintana Roo, Mexico. Dry-land fallow, although probably very important as a Maya agrosystem, leaves no archaeological remains, so constructional inputs are based on land clearance time, estimated to range from 19 to 25 work days/ha. As the Maya moved from simple to sophisticated agrosystems, i.e. from dry-land fallow to terracing and raised fields, exponential increases in constructional costs were incurred. It is probable that the Maya therefore progressed from low cost to high cost systems as demand for production increased. Major population growth occurred in the Classic, A.D. 500 to 900-1000, and it is precisely during this time of high production demand that the largest number of terrace and drained field systems appear to have been operative.

## Resumen

Se tratan los problemas del cálculo de rendimientos de cultivos a partir de sistemas agrícolas aborígenes abandonados. La interpretación de esos sistemas puede basarse sobre inversiones estructurales. Se comparan inversiones estructurales de campos drenados con el aterrazamiento de laderas y la barbechera de tierras secas en las tierras bajas maya. Los campos drenados constituyen el esfuerzo más costoso y tienen que haber estado asociados con una alta demanda para la producción y una economía de uso de la tierra intensivo. Datos de estructuras para campos drenados se toman del Pantano Pulltrouser, Belice, donde se estima que la construcción demoró entre 833 y 3116 días de trabajo por hectárea. El aterrazamiento constituye otra agrotecnología del período clásico en el area maya. La construcción de terrazas puede calcularse en términos de metros lineales o de metros cúbicos, pero un factor importante en el control de la velocidad de construcción es la distancia que deben arrastrarse las piedras para la construcción. Se estiman velocidades de metro cúbico de 1,4m$^3$ por día de trabajo o de 267 días de trabajo/hectárea y velocidades de metro lineal de 105 a 481 días de trabajo/hectárea para la construcción de terrazas de base ancha en el area del Río Bec del sur de Quintana Roo, México. El barbecho de tierras secas, a pesar de tener probablemente gran importancia como un agrosistema maya, no deja huellas arqueológicas, de tal manera que las inversiones estructurales se basan sobre el tiempo de desmonte de terreno que se estima entre 19 y 25 días de trabajo/hectárea. A medida que los mayas avanzaron desde agrosistemas sencillos hasta aquéllos más sofisticados, es decir desde barbecho de tierra seca hasta el aterrazamiento y campos elevados, se contrajeron aumentos exponenciales en los gastos estructurales. Es probable por lo tanto que los mayas hayan progresado desde sistemas de bajo costo a sistemas de alto costo a medida que aumentaba la demanda para la producción. Un importante crecimiento en la población ocurrió en el Clásico, 500 a 900-1000 D.C., y fue precisamente durante esta época de alta demanda de producción que parecen haber funcionado el mayor número de sistemas de aterrazamiento y de campos drenados.

The maturation of the study of ancient aboriginal agriculture in the New World has extended to an explanatory phase (e.g. Sanders, Parsons and Santley 1979; Turner and Harrison 1978; Turner 1983). To place the various agrosystems into descriptive sequences of development and into explanatory constructs of those sequences requires fine-tuned information on such variables as the morphology and constructional costs of the system, the frequency of cropping, and so forth. Unfortunately, much of this information for ancient, abandoned systems of agriculture will remain speculative. For example, it is doubtful that direct evidence of cropping frequencies or yields associated with abandoned agrosystems will be found. In some cases, comparisons can be made with extant systems, particularly if the comparisons involve the same environmental conditons and cultivation practices. For instance, maize (*Zea mays*) cultivation on chinampas (raised/drained fields) in the Valley of Mexico may be a good analogy for relic systems of chinampas found in other central Mexican locales. However, many abandoned agrosystems in the New World, such as the raised or drained field systems found throughout the wetlands of South America and the Maya lowlands of Central America (e.g. Denevan 1970; 1981; Turner and Harrison 1981; 1983), do not have modern counterparts in similar environments. Interpretations of these systems must be based on a mix of comparative information and theory.

One such attempt involves the use of constructional inputs as a means of comparing agrosystems in general and of assessing the systems in terms of theories of landscape modification or land-use change (e.g. Boserup 1965; Brookfield 1972; Turner *et al*. 1978). For example, it has been argued that the ancient Maya utilized a sequence of agrotechnologies through time and that these technologies as indicators of agrosystems were related to economic and environmental conditions (Sanders 1973; Turner 1974; 1983; Turner and Harrison 1978; 1981; 1983). Initial practices emphasized fallow systems on dry lands and orchard-gardens (Netting 1974; Harrison and Turner 1978; Flannery 1981; Turner and Miksicek n.d.) because of the low levels of demand placed on agriculture and the efficiency of fallow systems in dry land environments. However, as levels of demand increased and dry land became scarce, the Maya ultimately shifted to intensive systems, such as terracing on dry lands and field raising on wetlands. The resulting agrosystem allowed demand to be met but at a high cost of labour. Unfortunately, this interpretation (and others) has assumed that the input-output characteristics of systems probably conformed to those implied in the theory.

This paper evaluates the "data" on inputs for the three major agrotechnologies. Raised or drained fields are compared to terracing and dry-land fallowing. The results indicate that field raising in wetlands was a much more costly endeavour than were the other two and that an agricultural emphasis on the use of both raised fields in wetlands and terraces on dry lands was associated with times of high demand for production and a land intensive economy.

## MAYA AGROSYSTEMS

### Raised-Field Systems.

The central Maya lowlands possess a variety of seasonal and permanent wetlands (Siemens 1978; 1981; 1983; Turner 1978), locally referred to as swamps, *bajos, alkaches*, and savannas. The Prehispanic Maya manipulated many of these wetlands for a subtle but sophisticated form of hydraulic, raised-field cultivation, especially in the low-lying sections of the region (Siemens and Puleston 1972; Turner 1978; Harrison 1977; 1978; Turner and Harrison 1981; 1983), but also perhaps in the central Peten as well (Adams, Culbert and Brown 1981). These features were used in conjunction with canals for drainage and to regulate water level relationships to fields, not unlike the chinampa system of the Basin of Mexico (e.g. Armillas 1971; Sanders, Parsons and Santley 1979).

It is important to recognize that many, if not most, wetland fields in the Maya area were raised -- the surface of the field was elevated relative to the natural or original surface. Moreover, water was not drained away from the field zone but was maintained in the adjacent canals. In this regard, the Maya fields conform to similar systems in the Old World (Denevan and Turner 1974). These studies indicate that past and current fields of this kind usually involve some raising -- the mere digging of canals or channels produces earth that must be put somewhere. Moreover, this raising provides multiple functions (Denevan and Turner 1974). In some cases water is, indeed, drained from the "rooting zone" of the field, but in numerous systems, such as those indicated, water is not drained from the field/canal complex. For these reasons, Denevan and Turner (1974) chose the use of a "non-functional" generic term, raised fields. In as much as the wetland features discussed here are raised and did not involve drainage of water from the field zone, the raised-field designation is retained. It should be noted that these features are part of that complex agrotechnology referred to as drained fields in this book.

Maya raised-field agriculture is the subject of a considerable and increasing literature and will not be discussed in detail here (Turner 1974; in press; Harrison 1977; Puleston 1977a; Belise et al. 1977; Turner and Harrison 1978; 1981; 1983; Kirke 1980; Antoine, Skarie and Bloom 1981; Friedel and Scarborough 1981). The dating of Maya fields is somewhat controversial. Incipient wetland field systems may have been used as early as about 1,000 B.C., but large-scale systems in bajos and swamps apparently are a Classic period phenomenon (A.D. 300-1,000).

Intensive excavations of and interdisciplinary work on relic fields at Pulltrouser Swamp, northern Belize, has identified at least three constructional forms (Turner and Harrison 1981; 1983). The most simplistic type is the channelized field. This type was apparently constructed by cutting canals from the wetland proper into the adjacent swamp margin. The amount of raising of the intervening field is thought to be minimal. As suggested by Siemens (1981), the channelized field may represent the oldest or prototype of the true raised field. The latter were constructed within the wetlands and, at Pulltrouser, were apparently built in at least two ways. The first way (raised-field type A) involved the removal of most, if not all of the swamp top-soil, the excavation of about one metre deep canals into the underlying unconsolidated limestone, the creation of about one metre high platforms between the canals, and the placement of top soil on the platform. The platform itself was made of excavated canal material with other, perhaps transported fill, materials added (Turner and Harrison 1981; 1983). Raised field type A tends to be located adjacent to the dryland-wetland border. The second constructional form (raised-field type B) involved digging about one metre deep canals through the surface muck of the swamp into clay sediments. This material was used with that from the adjacent dryland to create one metre high platforms among the canals. Type B is found in the deeper or interior swamp segments and apparently did not involve the removal of the swamp top soil before creating the platform. Indeed, this soil remains buried under the fields.

Work-rate estimates for raised-field construction, particularly with traditional tools, are variable. This fact is not surprising given that detailed observations or experimentations on construction rates are few. Moreover, as demonstrated by the large amount of excavation data collected by the United Nations (1961), work rates vary considerably depending on the substance of excavation, the contractional arrangement of the workers, and the tools used. These observations have been confirmed in various studies. Wilken (1979; personal communication) has found a work rate of 0.83 $m^3$/hour (5.0 $m^3$/day) for one worker constructing a channelized/drained field in a peaty substance in Tlaxcala, Mexico. The same workman's rate dropped to 0.40 $m^3$/hour (2.4 $m^3$/day) when earth was encountered. Generally, the denser, wetter, or stickier the material, the slower the work rate. Experiments on simplistic raised-field construction in clays in Veracruz, Mexico, produced an apparent work rate of only 0.18 $m^3$/hour (1.1 $m^3$/day) (Gomez-Pompa *et al*. 1981). This calculation is somewhat questionable because the volume of construction is not clear. Experiments at Albion Island, Belize, produced variable rates depending on a contract by the hour or by the area to be completed (Puleston 1978; 1977b); the largest figure was 0.54 $m^3$/hour (3.2 $m^3$/day) for excavating and hauling a dry, unconsolidated, gypsum rich limestone (see Denevan 1981: Table 2:1). Erasmus (1965) demonstrated that a rate of 0.52 $m^3$/hour (3.1 $m^3$/day) can be achieved with digging sticks in limestone materials in the Maya area. In contrast, experiments in excavating chalk (limestone) in England with non-metal tools produced a rate of only 0.14 $m^3$/hour (0.8 $m^3$/day) (Ashbee and Cornwall 1961: 1331).

Here, work rates ranging from 0.40-0.83 $m^3$/hour (Table 2:2) have been used in order to err on the conservative side. The wet, clay-rich materials excavated at Pulltrouser Swamp probably could not have been excavated at rates higher than this range; indeed our observations suggest a rate more in line with the results from England and Veracruz, about 0.20 $m^3$/day. Utilizing the field dimensions and morphologies at Pulltrouser Swamp and "conservative" work rates, the range in work days per hectare is calculated at 833 to 3,166. If a lower work rate is used, following Gomez-Pompa and colleagues (1981) and Ashbee and Cornwall (1961), construction per hectare could be as high as 3,833 days. At Pulltrouser Swamp, approximately 10 percent of the fields are of the channelized variety and 90 percent are either raised field types A or B. Taking the lowest (833 and 958 work-days/ha) and highest (1,666 and 3,166 work-days/ha) values for the two categories and adjusting for their distribution, construction for all types of fields at Pulltrouser ranges from 945-3,016 work days per hectare.

### Terrace Systems

The Prehispanic Maya constructed a variety of terrace types across large tracts of the interior ridgelands of the Yucatan peninsula, particularly in southern Quintana Roo (where some exist in proximity to raised fields) and

Table 2:1

Labour for Construction of Maya Wetland Fields at Pulltrouser Swamp, Northern Belize

(2nd Approximation)

| Field Type | Feature | Area/ha[1] ($m^2$) | Volume/ha[1] ($m^3$) | Work Rate[2] ($m^3$/day/worker) | Work Days/ha | Range of Work Days/ha |
|---|---|---|---|---|---|---|
| Channelized Field | Canals | 4000 | 4000 | 4.8[3]<br>2.4[3] | 833<br>1666 | 833-1666 |
| Raised Field Type A | Canals | 4000 | 4000 | 4.8<br>2.4 | 833<br>1666 | |
| | Field<br>a. Removal of top soil | 6000 | 3000[4] | 4.8<br>2.4 | 625<br>1250 | 1583-3166 |
| | b. Platform construction | 6000 | 600[5] | 4.8<br>2.4 | 125<br>250 | |
| Type B | Canals | 4000 | 4000 | 4.8<br>2.4 | 833<br>1666 | |
| | Field | 6000 | 600[5] | 4.8<br>2.4 | 125<br>250 | 958-1916 |
| Type B | Canals | 4000 | 4000 | 1.2[6] | 3333 | |
| | Field | 6000 | 600[5] | 1.2 | 500 | 3833 |

[1]Based on mapping of Pulltrouser system and on recording that canals average one meter in depth and fields, one meter in height.

[2]Assumes 6 hour work days. Wilken (1979) and Denevan (1981) agree that a 5 hour effective work day may be more appropriate. Use of that figure would increase estimates in this table.

[3]Rates taken from Wilken (1979). Also see Denevan (1981).

[4]The original swamp top soil is estimated to have been 50 cm in depth.

[5]The platform took some labour to shape, but this figure is not known. Our evidence suggests that some non-swamp material was used in construction. Allowing for the expansion of the excavated canal material (clays), it is assumed that 10 per cent additional materials were used.

[6]Work rate consistent with the evidence from Gomez-Pompa _et al_. (1981) and Ashbee and Cornwall (1961).

Table 2:2

Labour for Maya Terrace Construction, Rio Bec Region, Yucatan Peninsula

(2nd Approximation)

| Location[1] | Area Mapped (ha) | Number of Terraces[2] | Linear Meters | Linear Meters of Terraces/ha | Construction Rate (Lm/day)[3] | Worker Days/ha[4] | Cubic Meters[5] | $M^3$/ha | Construction Rate ($M^3$/day)[6] | Worker Days/ha[4] |
|---|---|---|---|---|---|---|---|---|---|---|
| Site 1, 12.7 km east | 0.7 | 8 | 402.5 | 575.0 | 4.0 - 0.5 | 143-1150 | 402.5 | 575.0 | 1.4 | 410 |
| Site 2, 0.5 km west | 1.0 | 6 | 525.0 | 525.0 | 4.0 - 0.5 | 131-1050 | 525.0 | 525.0 | 1.4 | 375 |
| Site 10, 7.6 km east | 0.4 | 6 | 207.9 | 519.7 | 4.0 - 0.5 | 129-1039 | 207.9 | 519.7 | 1.4 | 371 |
| Site 12, 15.6 km east | 3.1 | 11 | 1428.9 | 460.9 | 4.0 - 0.5 | 115-921 | 1428.9 | 460.9 | 1.4 | 115 |
| Site 14, 34.7 km east | 6.5 | 11 | 1185.0 | 182.3 | 4.0 - 0.5 | 45-364 | 1185.0 | 182.3 | 1.4 | 130 |
| Site 16, 45.7 km east | 1.0 | 5 | 282.0 | 282.0 | 4.0 - 0.5 | 70-564 | 282.0 | 282.0 | 1.4 | 201 |
| Average | | | | 424.15 | | 105.5-848.0 | 671.9 | 424.15 | | 267 |

[1]Site designation refers to the distance from Xpujil, Campeche.

[2]Number does not include other stone features on slopes, such as boundary walls, walkways, or windbreaks.

[3]From Wilken (1979). Rates depend on several variables, the most significant of which is distance to source of rocks or construction materials

[4]Assuming that 6 hours is limit to daily output at the rates provided.

[5]Assuming a typical depth and width of one meter for intact terrace walls.

[6]Data from G. Wilken (personal communication) on construction rates of stone wall - rubble fill wier terraces (<u>trincheras</u>) in northern Mexico if no hauling of materials is required.

Campeche, Mexico, and in the Vaca Plateau of Belize. Others have been reported from central Peten, Guatemala, (Blom 1946; Turner 1974; Don Rice, personal communication) and in the savanna of southern Peten (Rice and Rice 1979). These features occur on all types of slopes dominated by limestone soils and commonly are associated with boundary walls, windbreaks, and other features. Maya terracing has been discussed in detail elsewhere (Turner 1974; 1978; 1980; 1983; in press; Turner and Harrison 1978; 1979; Healy, van Waarden and Anderson 1980; Rice and Rice 1979; also see Puleston 1978; Sanders 1979). It appears that true terracing was a Classic period (A.D. 300-1,000) agrotechnology in the Maya area.

At least two types of terraces, channel bottom weirs and linear sloping, dry field terraces were used. Most types were apparently constructed with stone walls of rather intricate design (Turner 1979; Healy, van Waarden and Anderson 1980). Large, broadbased walls of similar design to stone terraces elsewhere in Mesoamerica (e.g. Johnson n.d.; Willey et al. 1965: 575) are found in the Río Bec area of southeastern Campeche and southern Quintana Roo, Mexico. Walls are normally composed of two parts, a dry-laid front wall and a rubble buffer or fill of small "uniformly-sized" stones to the rear and under the front wall. A typical intact broadbase wall is about one metre high and one metre in width, including the rubble fill. The average linear metres of terraces per hectare from a sample of 6 mapped sites in the Rio Bec zone is 424 (Table 2:2). Based on the dimensions given, the volume is 671.88 cubic metres per hectare.

Two means are available for estimating constructional inputs, use of linear metres (Lm) or cubic metres ($m^3$). Wilken (1979) estimates that the linear metre construction rates of stone terraces range from 4.0 Lm to 0.5 Lm/day. The range results from the type and quality of construction and, more importantly, the distance of transporting the stone. Indeed, other works on excavation and hauling earth also indicate that the distance of hauling can be a major labour cost, involving diminishing returns to labour as distance increases (United Nations 1961). The linear metre approach suggests that Río Bec terrace construction ranged from about 105 to 848 days/ha, depending on hauling distance (Table 2:2). The volume figure, 1.4 $m^3$/day, produces a labour estimate of 267 days/ha for the broadbase terrace zone in question. The volume figure includes no rock hauling.

The cubic metre rate may be the most applicable of the work rates (but conservative) for the Río Bec broadbase terraces for several reasons. First, construction of these walls required more than piling stones. Examination of the fronting walls and, especially, the rubble fill indicates care in the selection or modification of the limestone used. Second, much of the stone for construction may have been available on the slopes or quarried close to the construction sites, requiring minimal hauling labour. Finally the volume construction rate is taken from trenches built with walls similar in design to those described here (Wilken, personal communication).

It is suggested that the broadbase terraces of the Rio Bec zone demonstrate an initial constructional input of at least 250 worker days per hectare (Table 2:2). This estimate does not include the other stone works found on the slopes and does not consider quarry or hauling activities.

Dry-land Fallow

It is recognized that the ancient Maya utilized various "fallow" practices that ranged in cropping frequencies from forest and bush rotations with slash-and-burn procedures to short and annual rotations, probably predicated on weeding, tilling, crop mixing, mulching and so forth (Turner 1979; 1983). Unfortunately for the student of prehistory, these systems can be pursued without implementing permanent landscape modifications, such that direct information about the system may never be recovered. For these reasons, direct evidence of dry-land fallow cultivation previous to ethnographic accounts is difficult to establish. Any possible indicator, such as pollen remains, can be interpreted in a number of ways (Turner 1983). Nonetheless, because of our general understanding of slash-and-burn and other types of fallow cultivation in the lowlands of Central America (e.g. Carter 1969; Nations and Nigh 1980; Wilk n.d.), these forms of agriculture are conceded to have a considerable history throughout the Maya area, particularly in those times and at those locales where land pressures were low.

Land clearance is the dry-land fallow counterpart to constructional inputs for terrace and raised-field systems. And, depending on vegetation, this input ranges from about 113 to 151 work hours (or 18.8 to 25.1 work days) per hectare (Carter 1969; 49; Wilk n.d.) throughout the Maya area. Since short-fallow cultivation, if practiced in the area, is not well documented, estimates of annual land clearance inputs are not known. Other inputs, such as tillage and weeding, are common to each of the three agrosystems in question and are best considered as procedural, not constructional inputs.

In principle, the ancient Maya could have pursued dry-land fallow agriculture throughout large segments of virtually every region of the Maya lowlands. Evidence of burning and forest destruction during early Formative or Preclassic times (ca. 2,000 B.C.) is suggestive that this agrosystem was used during the initial stages of cultivation in the region (Hammond and Miksicek 1981), although other agrosystems, such as orchard-gardens, may also be as early. More importantly constructional costs were limited to land clearance, and were low, particularly if land were cultivated several years in a row so that

| System | Constructional Inputs[1] (work/days/ha.) | Chronology of Use |
|---|---|---|
| Dry-land fallow cultivation | 19 - 25 | 2,000 B.C. (or earlier) to present |
| Terrace cultivation | 105.5 - 848 | A.D. 300 - 1,000 |
| Raised-field[2] (wetland) cultivation | 833 - 3,166 | About 1,000 B.C. - A.D. 1,000 |

1 One adult, male labourer working 6 hours per day

2 Incipient stages of the system may be earlier; its large-scale use may be from about 300 B.C. and certainly from A.D. 500.

Table 2:3 Comparison of Constructional Inputs and Dates of Maya Agrosystems

repeated clearance could be minimized.

## Interpretations

However crude, these data and estimates indicate that exponential increases in constructinal costs were incurred by the Maya as they moved from dry-land fallow to terraces and raised-field cultivation. Assuming that each "field" of cultivation was completely constructed before its use, then the constructional costs per hectare were about 19-25, 105-482, and 833-3,166 days of labour, respectively (Table 2:3). This information and the direct dating of the systems support the broader aspects of the agricultural growth model proposed at the outset of this essay. Holding constant environmental and other intervening variables, the ancient Maya progressed from low cost to high cost agriculture, based on initial construction inputs, as demands for production increased.

The Maya were cultivating maize before 2,000 B.C. Since no agrotechnologies have been dated to this early time, it is presumed that dry-land fallow and orchard garden systems were the dominant forms of production. During the following 2,000-2,500 years, these systems proliferated across the Maya realm as settlements grew in numbers and size (Ashmore 1981). Moreover, it is presumed that the intensity of production was increased when and where needed. In some areas, particularly where wetlands constituted a majority of a "land unit", increasing demand may have triggered experimentation with channelized fields as early as 1,000 B.C. (Siemens 1981). However, entry into major field raising in wetlands was probably a late Preclassic (300 B.C.-0) or later phenomenon, as indicated at Pulltrouser Swamp and Bajo Morocoy. The late Preclassic was a time of major population growth throughout much of the Maya area. From the late Preclassic period onward, the Maya probably experimented with a number of agrosystems and intensified dry-land production (Sanders 1973). Major population growth/numbers were obtained from about A.D. 500 to 900-1,000, the Classic period. It is precisely during this time of high production demands and growing land scarcity that the largest number of terrace and raised field systems appear to have been operative (Turner 1974; 1983; Turner and Harrison 1981; 1983; Healy *et al*. 1980). After A.D. 900-1,000 the Maya went through a societal transformation and a major reduction in population throughout much of their realm (Ashmore 1981). As demands for agricultural production decreased, high cost systems were abandoned and dry land fallow systems returned to dominance.

The obvious exception to this scenario is the apparent contradiction of the construction costs of terraces and raised fields and their incipient use (Table 2:3). If raised fields are so much more costly than terracing, then why would they appear earlier than terracing? It must be remembered that the large-scale of raised fields may be contemporaneous with terrace use. Raised fields dated to 1,000 B.C. are limited to Albion Island, Belize, and some questions remain about this information (Turner in press). More significantly, however, relaxation of environmental and population uniformity assumptions explains the possible use of wetlands before dry-land terracing. The evidence for Preclassic field raising is related to northern Belize, a region in which wetlands may equal dry lands in areal magnitude. It is also a region that was apparently occupied early in Maya history (Ashmore 1981; Hammond and Miksicek 1981). The result of these factors may have led to land pressures at times earlier than elsewhere in the Maya area, and hence, a need to move into costly raised-field agriculture. In contrast, the terraced areas of the Río Bec region are dominated by dry lands and did not incur similar levels of population pressure as did northern Belize until the Classic period. Regardless of these issues, field raising in northern Belize did not take place previous to or coevally with the development of dry-land fallow systems. The dry lands surrounding Pulltrouser Swamp show a long occupational history and were apparently deforested before field building in the swamp (Turner and Harrison 1983).

## Bibliography


ADAMS, R.E.W., W.E. BROWN, and T.P. CULBERT 1981, Radar mapping, archaeology, and ancient Maya land use, Science 213, (4515), 1457-1463.

ANTOINE, P.P., R.L. SKARIE, and P.R. BLOOM 1981, The origin of raised fields near San Antonio, Belize: an alternative hypothesis, In, Maya Subsistence: Studies in Memory of Dennis E. Puleston, edited by K.V. Flannery, Academic Press, New York 227-236.

ARMILLAS, P. 1971, Gardens on swamps, Science 174, 653-661.

ASHBEE, P. and I.W. CORNWALL 1961, An experiment in field archaeology, Antiquity 35, 129-134.

ASHMORE, W. 1981 (editor), Lowland Maya Settlement Patterns, University of New Mexico Press, Albuquerque.

BELISLE, J., M. SAID, and S. ASSAD, 1977 (editors), Journal of Belizean Affairs 5, Special Issue: The Rio Hondo Project and investigations of the Maya of Northern Belize, Belize City, St. Johns College.

BLOM, F. 1946, Apuntes sobre los ingenieros Mayas, Irrigacion en Mexico 27, 5-16.

BOSERUP, E. 1965, Conditions of Agricultural Growth, Aldine Press, Chicago, 124pp.

BROOKFIELD, H.C. 1962, Local study and comparative method: an example from Central New Guinea, <u>Annals Association of American Geographers 52</u>, 242-254.

BROOKFIELD, H.C. 1972, Intensification and disintensification in Pacific agriculture: A theoretical approach, <u>Pacific Viewpoint 13</u>, 30-84.

CARTER, W.E. 1969, <u>New Lands and Old Traditions: Kekchi Cultivators in the Guatemalan Lowlands</u>, Latin American Monographs No. 6, University of Flordia Press, Gainesville, 153pp.

DENEVAN, W.M. 1970, Aboriginal drained-field cultivation in the Americas, <u>Science 196</u>, 647-654.

DENEVAN, W.M. 1981, Hydraulic agriculture in the American tropics: forms, measures and recent research, <u>In, Maya Subsistence, Studies in Memory of Dennis E. Puleston</u>, edited by K.V. Flannery, Academic Press, New York, 181-203.

DENEVAN, W.M. and B.L. TURNER II 1974, Forms, fucntions and associations of raised fields in the Old World tropics, <u>Journal Tropical Geography 39</u>, 24-33.

ERASMUS, C.J. 1965, Monument building: some field experiments, <u>Southwestern Journal of Anthropology 21</u>, 277-301.

FRIEDEL, D.A. and V. SCARBOROUGH 1981, Subsistence, trade and development of the coastal Maya, <u>In, Maya Subsistence: Studies in Memory of Dennis E. Puleston</u>, edited by K.V. Flannery, Academic Press, New York, 131-155.

FLANNERY, K.V. 1981 (editor), <u>Studies in Memory of Dennis E. Puleston</u>, Academic Press, New York, 368.

GOMEZ-POMPA, A., H.L. MORALES, E. JIMENEZ and J. JIMENEZ 1981, Experiences in traditional hydraulic agriculture, <u>In, Maya Subsistence: Studies in Memory of Dennis E. Puleston</u>, edited by K.V. Flannery, Academic Press, New York, 327-342.

HAMMOND, N. and C. MIKSICEK 1981, Ecological economy of a formative Maya site at Cuello, Belize, <u>Journal of Field Archaeology 8</u>, 260-269.

HARRISON, P.D. 1977, The rise of the <u>Bajos</u> and the fall of the Maya, <u>In, Social Process in Maya Prehistory</u>, edited by N. Hammond, Academic Press, London, 469-508.

HARRISON, P.D. 1978, <u>Bajos</u> revisited: visual evidence for one system of agriculture, <u>In, Pre-Hispanic Maya Agriculture</u>, edited by P.D. Harrison and B.L. Turner II, University of New Mexico Press, Albuquerque, 247-253.

HEALY, P.F., C. van WAARDEN and T.J. ANDERSON 1980, Nueva evidencia de antiquas terrazas Mayas en Belize, America Indigena 40, 773-796.

JOHNSON, K. 1977, "Do as the Land Bids", A Study of Otimi Resource-Use on the Eve of Irrigation, Ph.D. dissertation, Clark University, 533pp.

KIRKE, C.M. ST.G. 1980, Prehistoric agriculture in the Belize River Valley, World Archaeology 2, 281-287.

NATIONS, J.D. and R.B. NIGH 1980, The evolutionary potential of Lacadon Maya sustained-yield tropical forest agriculture, Journal of Anthropological Research 36, 1-29.

NETTING, R.Mc. 1974, Maya subsistence: mythologies, analogies, possibilities, In, The Origins of Maya Civilizations, edited by R.E.W. Adams, University of New Mexico Press, Albuquerque, 299-333.

PULESTON, D.E. 1977a, The art and archaeology of hydraulic agriculture in the Maya Lowlands, In, Social Process in Maya Prehistory, edited by N. Hammond, Academic Press, London, 449-475.

PULESTON, D.E. 1977b, Experiments in prehistoric raised field agriculture: learning from the past, Journal of Belizean Affairs 5, 36-43.

PULESTON, D.E. 1978, Terracing, raised fields, and tree cropping in the Maya Lowlands: a new perspective in the geography of power, In, Pre-Hispanic Maya Agriculture, edited by P.D. Harrison and B.L. Turner II, University of New Mexico Press, Albuquerque, 225-246.

RICE, P.M. and D.S. RICE 1979, Home on the range - aboriginal Maya settlement in the Central Peten Savannas, Archaeology 32, 16-25.

SANDERS, W.T. 1973, The cultural ecology of the Lowland Maya: a re-evaluation, In, The Classic Maya Collapse, edited by T.P. Culbert, University of New Mexico Press, Albuquerque, 325-365.

SANDERS, W.T. 1979, The jolly green giant in tenth century Yucatan, or fact and fancy in Classic Maya agriculture, Reviews in Anthropology 6, 493-506.

SANDERS, W.T., J.R. PARSONS and R.S. SANTLEY 1979, The Basin of Mexico: Ecological Processes in the Evolution of a Civilization, Academic Press, New York, 561pp.

SIEMENS, A.H. 1978, Karst and the Prehispanic Maya in the Southern Lowlands, In, Pre-Hispsanic Maya Agriculture, edited by P.D. Harrison and B.L. Turner II, University of New Mexico Press, Albuquerque, 117-143.

SIEMENS, A.H. 1981, Prehispanic agricultural use of the wetlands of Northern Belize, In, Maya Subsistence. Studies in Memory of Dennis E. Puleston, edited by K.V. Flannery, Academic Press, New York, 205-225.

SIEMENS, A.H. 1983, Oriented raised fields in Central Veracruz, American Antiquity 48, 85-102.

SIEMENS, A.H. and D.E. PULESTON 1972, Ridged fields and associated features in Southern Campeche: new perspectives on the Lowland Maya, American Antiquity 37, 228-229.

TURNER, B.L. II 1974, Prehistoric intensive agriculture in the Mayan Lowlands, Science 185, 118-24.

TURNER, B.L. II 1978, Ancient agricultural land use in the Central Maya Lowlands, In, Pre-Hispanic Maya Agriculture, edited by P.D. Harrison and B.L. Turner II, University of New Mexico Press, Albuquerque, 163-182.

TURNER, B.L. II 1979, Prehispanic terracing in the Central Maya Lowlands: problems of agricultural intensification, In, Maya Archaeology and Ethnology, edited by N. Hammond and G.R. Willey, University of Texas Press, Austin, 101-115.

TURNER, B.L. II 1980, La agricultura intensive de trabajo en las Tierras Mayas, American Indigena 40, 653-670.

TURNER, B.L. II 1983, Once Beneath the Forest: Prehistoric Terracing in the Río Bec Region of the Maya Lowlands, Dellplain Series in Geography, Westview Press, Boulder, 209pp.

TURNER, B.L. II (in press), Comparative analysis between agro-systems in the Basin of Mexico and Central Maya Lowlands: formative to the Classic Maya collapse, In, Interdisciplinary Approaches to the Study of Mesoamerican Highland-Lowland Interaction, A conference at Dumbarton Oaks, Washington, D.C. October 18-19, 1980, edited by Arthur Miller, Washington D.C., Dumbarton Oaks Research, Library and Collection, Trustees to Harvard University (publication in 1985).

TURNER, B.L. II and P.D. HARRISON 1978, Implications from agriculture for Maya prehistory, In, Pre-Hispanic Maya Agriculture, edited by P.D. Harrison and B.L. Turner II, University of New Mexico Press, Albuquerque, 337-373.

TURNER, B.L. II and P.D. HARRISON 1979, Comment on W.T. Sander's review of Pre-Hispanic Maya Agriculture, Reviews in Anthropology 6, 544-55.

TURNER, B.L. II and P.D. HARRISON 1981, Prehistoric raised-field agriculture in the Maya Lowlands, Science 213, 399-405.

TURNER, B.L. II and P.D. HARRISON, (editors) 1983, Pulltrouser Swamp: Ancient Maya Habitat, Agriculture and Settlement in Northern Belize, University of Texas Press, Austin.

TURNER, B.L. II and W.C. JOHNSON 1979, A Maya dam in the Copan Valley, Honduras, American Antiquity 44, 299-305.

TURNER, B.L. II and C. MIKSICEK n.d. Economic species associated with prehistoric agriculture in the Maya Lowlands, Paper presented at the Symposium on History and Ecology of Crops and Cropping Systems in the Americas, 13th International Botanical Congress, Sydney, 1981.

UNITED NATIONS 1961, Earth Moving by Manual Labour and Machines, Bangkok, Economic Commission for Asia and the Far East, 114pp.

WILK, R.R. 1981, Agriculture, Ecology and Domestic Organization Among the Kekchi Maya, Ph.D. dissertation, University of Arizona.

WILKEN, G.C. 1979, Traditional slope management: an analytical approach, In, Hill Lands: Proceedings of an International Symposium, edited by J. Luchok, University Books, Morgantown, West Virginia, 416-422.

WILLEY, G.R., W.R. BULLARD, Jr., J.B. GLASS and J.C. GIFFORD 1965, Prehistoric Maya settlement in the Belize Valley, Papers of the Peabody Museum of Archaeology and Ethnology 54, 1-589.

Chapter 3

# MODELLING PRE-HISPANIC HYDROAGRICULTURE ON LEVEE BACKSLOPES IN NORTHERN VERACRUZ, MEXICO

Alfred H. Siemens

## Abstract

Parallel evidence from Pre-Hispanic and present land use practices in the riverine wetlands of northern Veracruz indicates a tradition of agricultural adaptations to the annual water level fluctuations of the floodplain environments. Both past and present cropping schemes are strongly determined by the amplitude of the annual water level and both Pre-Hispanic settlement remains and contemporary settlements are found in these positions. Permanent agriculture is possible here. The areas of the floodplain subject to seasonal inundation are presently occupied by pastures which from the air show vestiges of ancient canals and planting platforms. Based on both contemporary and Pre-Hispanic evidence a hydroagricultural system for the region is proposed which divides the wetland environment into four segments: levee top, upper and middle backslope, lower backslope and backswamps. The canals on the backslope are seen as aiding the rapid drainage of water at the end of the wet season which allows an earlier beginning to cultivation than would otherwise be possible. It is also likely that the canals were used to impound water and prolong dry season cropping. The annual cropping cycle may therefore be seen as beginning on the levee tops and ending on the lowest fields in the backswamp. This type of agricultural system, still practiced in northern Veracruz and other areas, is a move towards agricultural intensification and risk reduction.

## Resumen

Existe evidencia paralela de prácticas del uso de la tierra prehispánico y actual en las tierras húmedas fluviales del norte de Veracruz que indica una tradición de adaptaciones agrícolas a las fluctuaciones anuales del nivel del agua dentro del medio ambiente de la llanura de inundación. Tanto los proyectos de cultivos en el pasado como actualmente estan en gran medida determinados por la amplitud de la variación anual del nivel del agua - la parte superior de los terraplenes se encuentra normalmente sobre el nivel alto anual y tanto los restos de los asentamientos prehispánicos como los asentamientos actuales se encuentran en estas posiciones. Aquí es posible practicar la agricultura a través de todo el año. Las areas del llano de inundación que estan sujetas a inundaciones estacionales se encuentran actualmente ocupadas por pastizales que desde el aire presentan vestigios de antiguos canales y plataformas de plantación. Basado sobre evidencia tanto contemporánea como prehispánica se propone un sistema hidroagrícola para la región que divide el medio ambiente del suelo húmedo en cuatro partes: cumbre de terraplén, parte superior y media de ladera, ladera inferior y pantanos. Los canales en las laderas contribuyen a un rápido drenaje de agua al fin de la estación lluviosa, lo cual permite comenzar con las siembras antes de lo que sería posible de otra manera. También es probable que los canales se hayan usado para embalsar el agua y así prolongar los cultivos de la estación seca. De tal manera que el ciclo anual de siembras puede verse como comenzando en las cumbres de los terraplenes y terminando en los campos más bajos del pantano. Este tipo de sistema agrícola, aun practicado en el norte de Veracruz y otras areas, significa un paso hacia la intensificación agrícola y la reducción del riesgo.

In air surveys over Northern Veracruz and related vertical air photo analysis from 1977 to 1982 the remains of over two dozen substantial complexes of canals and planting platforms have been detected in the riverine wetlands of the region (Figure 3:1; Siemens 1982). A characteristic assemblage of features recurs, as well as certain environmental relationships and a micro-altitudinal zonation. From all of this one may deduce an agricultural strategy, a way of utilising the soil and water resources of terrain subject to seasonal inundation, which comes down, in large part, to the problem of reducing risk. This strategy seems similar to that represented by past and present agriculture on a <u>terra firma</u>/wetland boundary zone recently examined in Central Veracruz and instances of the dry season cultivation of levee backslopes reported from numerous locations in Mesoamerica (Siemens 1983c). It is closely related to the Prehispanic agricultural use of wetlands already described in general terms in other contexts (e.g. Siemens 1983b) which may be said to resemble the <u>chinampas</u> of Central Mexico but must not be equated with them.

Modelling the assemblage on the basis of available data, i.e. the results of a visual analysis of wetland landscapes as well as a limited ground reconnaissance and ethnographic enquiry, should be useful in a number of respects. It should enhance sensitivity to micro-environmental variations, as necessary in the discussion of agriculture around and within wetlands as in any other context. It helps the observer to link agricultural activities micro-altitudinally, which the farmers of the lowlands have always been constrained to do themselves. Furthermore, it accommodates what is known or may reasonably be imputed about the construction of the canals and planting platforms. Basically, it may help in making sense of morphological variations, which constitute the most accessible evidence on the nature of the system, and improve the generalisation that can be made about a phenomenon that occurs in various environmental, cultural and chronological contexts. At the very least it can serve as a foil for further discussion. Detailed further investigation is planned, including archaeological excavations and sedimentological sampling, as well as topographic mapping and air reconnaissance at various times of the year.

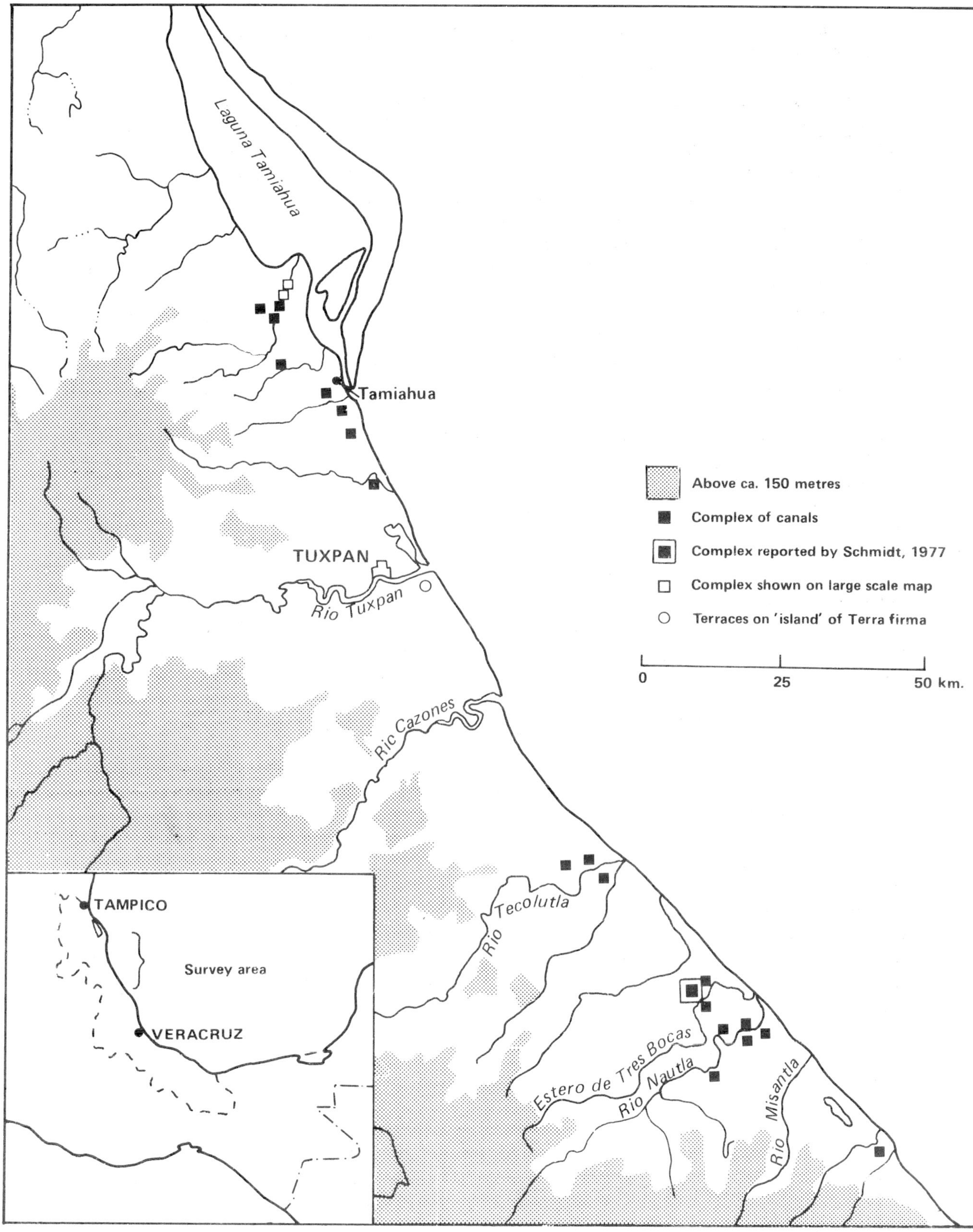


Figure 3:1 Complexes of canals and planting platforms found in Northern Veracruz

It is necessary to clarify several semantic points before proceeding. Some of the terms commonly used in the discussion of this sort of agriculture have become inaccurate, in particular the modifiers *raised* and *drained*, attached to the vaguer and still useful *fields*. Both excavations and build-up are involved in the construction of the features that pattern many Mesoamerican wetlands. In some situations, particularly on the upper edges of the very gentle saucer-like basin of each wetland, the facilitation of the drainage function seems to have been pre-eminent. However, some build-up, if only from the spoils of canal excavation, will have been involved as well. Nearer the centre of the wetland, build-up would seem to have been more important, perhaps necessitating transport of material from *terra firma*. Moreover, the same web of canals may be envisioned to have served both to conduct water away from the wetland fields, especially during the later stages of the wet season and the beginning of the dry season, as well as to store it as a source for irrigation later in the dry season. Neither *drained* nor *raised* is thus entirely satisfactory. One or the other may be needed occasionally as a code word representing the whole phenomenon, in which case one could argue about as well as for the one as the other. In order, therefore, not to prejudge questions of construction and function, the fairly neutral terms, *canals* and *planting platforms* will be used here repeatedly.

## Environmental Factors Facilitating Agricultural Intensification in Northern Veracruz

The field systems are found on Quaternary alluvium laid down between hill land and the sea. The hills are underlain mostly by horizontally bedded Tertiary marine sediments (Carta Geologica de Mexico 1977). Viewed from the air they often show a stepped surface, which at first sight suggests vestigial terraces. At several roadcuts it was possible to verify that they were merely the surface expression of resistant strata. The boundary between hill land and recent alluvium is an extremely crenelated affair. The alluvium itself presents the usual floodplain sequence: levee, gentle backslope and backswamp. Where streams meander extensively these features are often occluded, creating shallow basins.

The average annual precipitation in the coastal lowlands of Northern Veracruz is 1500-2000 mm around Nautla, 1200-1500 mm around Tuxpan and Tamiahua (Garcia 1970). Less than 5 percent of the total normally falls during the dry season, i.e. between November and April. This is a principal reason why dry season agriculture, often called *tonalmil*, on hill land is quite tenuous; it usually yields a good deal less than the wet season crop on the same land. Nevertheless, it is widely practised and indeed increasing in frequency (Camou 1981). Under such circumstances, currently as in the distant past, nearby wetlands represent an increasingly attractive resource.

Regarding natural plant cover, Coastal Northern Veracruz is represented as generally able to sustain high tropical rain forest (Rzedowski 1978). Very little of this remains; such secondary forest growth as is still to be found is rapidly being cleared as well. The most obtrusive aspect of the current vegetation, especially in the wetlands, is the predominance of recently introduced grasses. As it happens, these allow the vestiges of ancient land use to show through much better than the natural plant cover.

For clues to the behaviour of the intermittent streams that drain the wetlands and the nature of flooding, the key aspects of the natural environment in this discussion, one is dependent mainly on direct observation and on information from local inhabitants. It is apparent that water rises and falls in the backswamps on the floodplain of Northern Veracruz to similar levels as it does in the relevant main stream channels. Moreover, there are indications that the present fluctuation is not too different from what it was during the operation of the ancient field systems, whenever that may turn out to have been. On many levees the downslope limits of the remains of Prehispanic settlements coincide roughly with the limits of the zone occupied by modern farmsteads, roads and villges. This cultural strand-line, in turn, often coincides with both the upper limits of the vestiges of canalization and current flooding. This strand-line is also at approximately the level of high water in the stream bed. Nevertheless, this is subject to fine tuning once the stratigraphy of the canals and planting platforms can be examined. On the vertical air photo coverage of some of the wetlands earlier, more constricted, topographically conformal boundaries are apparent. This is a hint that water levels have risen and that the wetlands have expanded slightly in the not too distant past.

Measurements are available from several hydrometric stations on the larger through-flowing streams such as the Nautla, along which several field investigations have been undertaken (Table 3:1). It is evident that the yearly fluctuation in water level in the main stream and hence in the floodplain of the Nautla is limited on the average to something over two metres, which is slightly higher but still quite similar to the type of regime found to have been manageable for Prehispanic wetland agriculture elsewhere in the Mesoamerican tropical lowlands, in particular along the karstic rivers on the two flanks of the Yucatan Peninsula (Siemens 1978).

Along the Estero Tancochin, a region just to the north of Tamiahua, where a field investigation was undertaken as well, the annual fluctuation in the backswamps is reported by local informants to be from three to five metres. This is appreciably higher than the measured mean fluctuation

| Year | Max. Level/metres | Min. level/metres | Difference |
|---|---|---|---|
| 1952 | 2.14 | 1.54 | 0.6 |
| 1953 | 4.53 | 1.46 | 3.07 |
| 1954 | 5.10 | 1.41 | 3.69 |
| 1955 | 6.46 | 1.40 | 5.06 |
| 1956 | 3.25 | 1.55 | 1.70 |
| 1957 | 3.26 | 1.45 | 1.81 |
| 1958 | 3.80 | 1.40 | 2.40 |
| 1959 | 3.18 | 1.34 | 1.84 |
| 1960 | 2.56 | 1.27 | 1.29 |
| 1961 | 3.32 | 1.41 | 1.91 |
| 1962 | 3.23 | 1.30 | 1.02 |
| 1963 | 3.16 | 1.28 | 1.88 |
| 1964 | 2.84 | 1.30 | 1.54 |
| 1965 | 3.00 | 1.36 | 1.64 |
| 1966 | 4.80 | 1.36 | 3.44 |
| 1967 | 5.70 | 1.29 | 4.41 |
| 1968 | 2.56 | 1.21 | 1.35 |
| 1969 | 5.37 | 1.33 | 4.04 |

Arithmetic mean of annual difference: 2.37 metres

Extreme annual difference for period: 5.00 metres

(SRH 1970)

Table 3:1 Water levels at Martinez de la Torre on the Rio Nautla

along the Nautla, and yet here too the floodplain is extensively patterned.

The short run of data available for the Nautla also shows several maxima considerably above the mean and at least one quite extraordinary fluctuation (1955 - 5.06 metres above minimum level of that year). In a 15-year period the Candelaria also fluctuated twice to several times its mean (Siemens 1978). A major flood is reported for 1955 along the Tecolutla, the river to the north of the Nautla, along which numerous complexes of ancient fields were also found. Archaeological evidence indicates that such extraordinary floods have occurred along the Tecolutla since Middle Formative times (Wilkerson 1980).

A good deal remains to be done generally on flooding *vis-a-vis* the system of canals and planting platforms in the wetland in the lowlands. The yearly use of the water level, even though it might go near the tops of the platforms may well have been beneficent, depositing new sediments, killing pests and weeds, acting in effect like a short fallow. A flood of several metres above the normal, however might be expected to have currents and thus would seriously erode earthworks; it would also damage homes and gardens on the levee top and disturb cropping sequences generally throughout the wetland.

In general, the riverine wetlands of the lowlands have been attractive to farmers whenever they needed to go beyond what they could produce in wet season milpa on *terra firma* and to cattlemen from the beginning of the colonial period onwards. The soils are easily worked and potentially fertile. Water levels fluctuate, but such seasonal flooding as occurs can be taken in stride. The stored water is a valuable resource during the substantial dry seasons that recur throughout the region. The wetlands are also usually in close proximity to hill land suitable for wet season cultivation. They are, therefore, far from the empty, useless areas they have often been described to be, but rather may be seen as favoured locations and key segments of a population's sustaining area.

A number of Mesoamerican lowland regions with large expanses of wetland were given the aura of abundance by the chroniclers of contact. The Maya province of Acalan, around the floodplain of the Candelaria River in what is now the state of Campeche, is a notable example (Figure 3:1). The first Spanish expedition into that region found the provisions it desperately needed and later *encomenderos* obtained from it a wide range of food products as tribute (Scholes and Roys 1968). The exploitation of riverine wetlands may have been part of the reason for that productivity, but there is no direct reference to it. The region was virtually depopulated after Conquest and not reoccupied to any great extent until the middle of the twentieth century. Nevertheless, the floodplains of the Candelaria and its tributaries are webbed in many places with the vestiges of canals. Some of them will have served mainly transportational purposes; others enclosed planting platforms. They seem to have been in use as early as the Maya Late Formative or Early Classic, i.e. 300 B.C. to A.D. 300 (Siemens and Puleston 1972). Similar remains with intimations of congruous or even earlier dates were subsequently found in other lowland Maya locations, especially northern Belize and southern Quintana Roo (Turner 1974; Siemens 1977; Puleston 1977; Turner and Harrison 1983).

In 1977 the author's investigation of the Prehispanic use of wetlands was taken out of the Maya area into Central Veracruz. The coastal lowlands of this part of the state, wedged between the dunes and the onset of the hilly topography that rises gradually to the escarpment of the Sierra Madre Oriental, are pockmarked with numerous wetlands. Almost every one of these has been found to be scored by the remains of canals and planting platforms. The cluster of wetlands that take up much of the San Juan River basin, which is drained northward into the Antigua River, is now under intensive investigation. It is already apparent that most of the ancient field complexes of this basin, and indeed, the other wetlands of Central Veracruz as well, are oriented roughly in the sacred direction of Teotihuacan - $15^{o}$ 25' east of N (Siemens 1983a). This upland centre dominated much of Central Mexico during the first seven centuries after Christ and seems to have contributed a landscape-ordering principle to the coastal lowlands of Central Veracruz as well.

The Cortés expedition found the wetlands just behind what is now the port of Veracruz productive and indeed refreshing - a place of abundant food and rich greenery (Diaz, 1976; Wilkerson Chapter 4). Whether or not the planting platforms were in use at the time is not clear, but there is evidence for irrigation in the surroundings of Zempoala (Palerm 1953).

The population of this region was also severely reduced after Conquest; the wetlands reverted to forested wilderness and remained so throughout the colonial period and the nineteenth century. Ranchers drove their herds into the wetlands during the dry season (Sartorius 1961); the numerous half-wild cattle will have sought them out on their own. Travellers saw very little agriculture of any kind between Veracruz and Jalapa (Thomspon 1846). Alexander von Humboldt (1811) was impressed by the potential of the region - and its disuse. In the second half of the twentieth century the wetlands of Central Veracruz have once again become abundant with crops and pastures.

In 1981 the search for remnants of wetland agriculture was extended into Northern Veracruz. This investigation is still in its preliminary stages; the methods applied to date include repeated small plane reconnaissance, vertical air photo analysis and ground transecting. They have yielded evidence of extensive rectilinear patterning in the floodplains of the Nautla and Tecolutla Rivers as well as along a series of lesser streams to the south and north of Tamiahua (Figure 3:1).

Early post-Conquest ethnohistory offers some evocative imagery regarding abundance in Northern Veracruz. The Aztecs, it seems, believed their place of origin lay somewhere near the coast of Northern Veracruz (Duran 1964). They applied to these warm, moist lowlands, including the territory of the Huastecas and the Totonacas, the name Tonacatlalpan, the "land of food" (Straesser-Pean 1968). There had been plenty of water, the environment had offered food resources of many kinds and, most importantly, the ancestors had developed an excellent cropping system. The evidence from the wetlands substantiates this picture more and more.

## Several Specific Canal and Planting Platform Complexes

The most striking remains in Northern Veracruz are those to be found along the Estero de Tres Bocas, which joins the Nautla River near its mouth (Figures 3:1, 3:2 and 3:3). This complex is similar in a number of respects to others located along the same stream, as well as the Nautla and Tecolutla Rivers. It was reported by Schmidt in 1977, together with many useful observations on the relationship of the complex to research on ancient subsistence sytems in the Mesoamerican tropical lowlands (Schmidt 1977). Canals have also been reported from along the Tecolutla (Wilkerson 1979, 1980, Chapter 4).

The Estero may be seen in the lower left corner of Figure 3:2. The levee top, normally above the yearly high water level, offers the prime settlement sites of the region. The imposing mounds pertain to the ruin complex known as Santa Elena. The headquarters of the ranch that encloses it are just off the photograph to the left. Most of the natural forest cover of the levee tops is gone. The groups of palms in the pasture echo long human occupance and use (Rzedowski 1978). Plantations of bananas and citrus fruits appear in the foreground. Machinery is in the field in the lower right corner, preparing for or actually seeding a wet season crop of maize. The patterns left by machines are apparent in the bottom third of Figure 3:3 as well, where they represent a mature or already harvested dry season crop of maize. This is a mechanised form of the dry season crop traditionally known in the region as *tonalmil*. Both instances symbolise the effects of a policy of food production increase adopted at federal and state levels during the last years of the Lopez Portillo Presidency and known as the Sistema Alimentario Mexicano (SAM). In Veracruz it often meant the expansion of crop land at the expense of pasture, as illustrated. Such expansion was found to be expensive and questionable on other grounds; the policy has since lapsed.

Figure 3:2 The Santa Elena Site and remains of wetland agriculture north of the Estero de Tres Bocas

<u>Figure</u> <u>3:3</u> Detail of remains of canals and planting platforms north of the Santa Elena site

A boundary zone, a kind of cultural and hydrological strand-line, divides the reasonably secure levee tops from the lower land subject to seasonal inundation. It may be detected diagonally through the top right of Figure 3:2 and also through the top right of Figure 3:3 (the same four palm trees that appear in the top left corner of Figure 3:2 are indicated by an arrow in the top right corner of Figure 3:3). A similar zonation can be seen on the floodplains of many of the streams of Northern Veracruz.

Those parts of the floodplain behind the ruins of Santa Elena that are subject to seasonal inundation are occupied by pastures, except for the lowest portions, which may be occupied by hydrophytes or open water all year. Fences and the bold tonal differences induced by variations in pasture use dominate most floodplain landscapes of the region when seen from the air. Equally striking vestiges of canals and planting platforms show through the pastures in many places, as in Figures 3:2 and 3:3. The differences in tones reflects differences in vegetation. They are particularly noticeable towards the end of the wet season, when the contrast between the moisture in the former canals and platforms is the greatest. The pattern is roughly rectilinear; many of the rectangles on the upper margins of the zone of inundation are open towards the top of the levee. The topographic variation between the bottom of the canal vestiges and the tops of the adjacent fields is a matter of approximately 30 centimetres. From local informants it was apparent that the yearly fluctuation of the water level, i.e. the distance between the flood level in the wet season and the groundwater level in the dry season, was from two to three metres, which is similar to that measured along the Nautla River (Table 3:1). It is the amplitude of this fluctuation that is of particular interest here rather than water levels *vis-a-vis* the eroded and filled in remains *per se*.

On vertical air photographs covering the northern side of the floodplain of the Estero (CMA 1968) it is possible to distinguish the direction of drainage down the levee slope from the Santa Elena ruin complex, as well as low topographic passes through which the floodwaters recede into the lowest parts of the backswamp and eventually into the main stream. A control of drainage by means of simple barrage dams in the passes is entirely plausible. Some detailed topographic mapping and archaeological testing is planned in aid of this hypothesis, as well as other aspects of the model proposed in this paper.

A similar ecology of settlement and agriculture may be seen with a key addition several kilometres downstream from Santa Elena (Figure 3:4). The ruins of a Prehispanic centre and the houses of the present inhabitants are again arrayed along the top of the levee, together with extensive household gardens, ranch facilities and plantations of banana and coconut. The raked pattern of

Figure 3:4 Surroundings of the Estero de Tres Bocas several kilometres downstream from Santa Elena

ancient canalisation is apparent on the backslopes, with the intervening fields open on their upslope ends. A recent analogue is also apparent in the immediate foreground of the photograph. The backslope there is streaked with the remains of canals that probably pertained to a plantation antedating the mechanised <u>tonalmil</u>. Schmidt presents an illustration of this sort of canalisation in a banana plantation he visited in this same region (1977, p.56).

The patterns noticeable on the floodplains of the lower Nautla basin may be seen in numerous other locations to the north. Along the Tecolutla River, a complex is apparent just to the north of the town of Gutierrez Zamora, several more to the north and south of the river downstream from the town. The latter although not as striking visually as those along the Estero de Tres Bocas, illustrate a similar field pattern, location along a levee backslope and drainage by means of a main, canalised collector into the river at a point downstream from the complex (CMA, 1954, 1:35,000, f.1130, numbers 82-185).

The canals and fields sited along streams to the north and south of Tamiahua are represented by the segments out of the floodplain of the Estero de Tancochin shown in Figures 3:5 and 3:6. The canal networks around Tamiahua are not easily mapped in their entirety, nor are those near Santa Elena, for that matter, even though they appear strikingly clear at first sight. Discontinuities and ambiguous stretches are numerous. The vertical air photo coverage available for the surroundings of the Estero (Detenal 1976) allows the recognition of the main drainage channels out of the field complexes but only small portions of the narrower canals that define the field systems. The oblique air photographs taken during this investigation (e.g. Figure 3:5) show vastly more detail because they are larger in scale and benefit from widespread recent clearing and the planting of new pastures. However, the oblique view severely distorts all patterns. Consequently, canal networks have been mapped here in generalised terms.

The floodplain within the bends of the Estero is often occluded, i.e. it does not show the full levee backslope into a backswamp, but instead forms a shallow saucer-like depression patterned from side to side. The levee-top shown in the top right corner of Figure 3:4 is quite narrow; that on the other side (Figure 3:5) is wider and surmounted by a group of mounds. They are not as spectacular as those of Santa Elena, but similarly situated. Among them are the headquarters of a ranch, as well.

The canal vestiges (5-6m wide) define long narrow planting platforms (ca. 25m wide); many of the platforms near the boundary of the levee top and the zone of seasonal inundation are open up-slope.

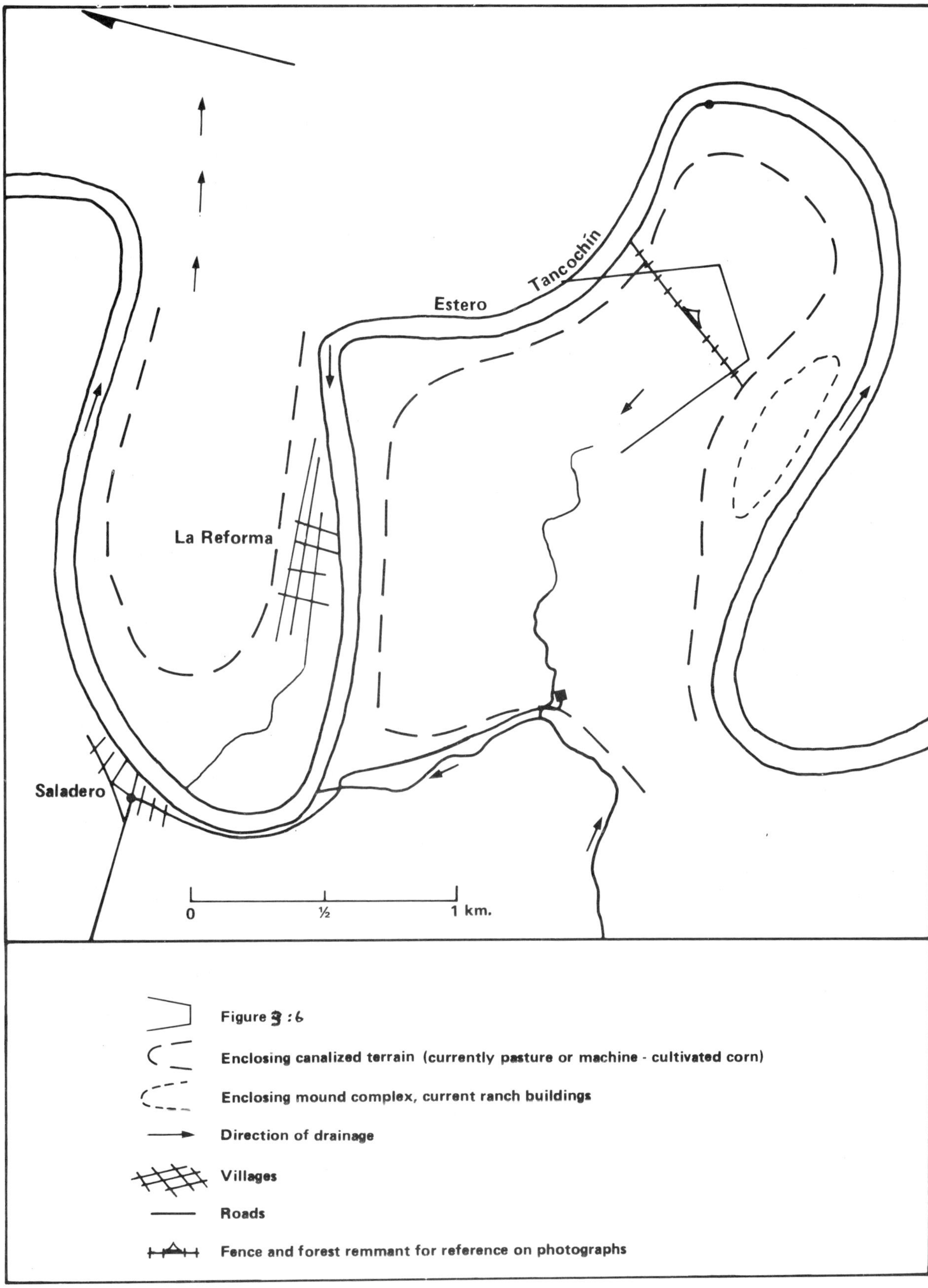


Figure 3:5 Environs of the Estero Tancochin

Ground investigations indicated that in October 1980 the floodplain had been inundated up to the margin of settlement on the levee tops. In late January 1981, the time at which the photograph (Figure 3:5) was taken, there was still water in the vestiges of the canals, which are only 10 to 20 centimetres deep. Maize had been seeded in machine-prepared soil on segments of the floodplain. By May of the same year the groundwater level was approximately 2 to 3 metres below the surface. The maize was tall now, its ears nearing maturity. It was apparent that there had been a total fluctuation in the water level over the last twelve-month period of 3 to 4 metres. In some years high-water comes several metres higher and floods the streets of the community of Saladero, which is laid out along the top of a levee.

It is not difficult to locate the main drainage channels out of the field complexes, as was indicated. It is clear that they were straightened for considerable stretches and one can easily locate topographic passes in this region as well, where the recession of floodwaters could have been retarded by damming. The overall form of many field complexes in the floodplains of Northern Veracruz is roughly oblong and parallel to the main stream, connected to it by a drainage channel at a point some distance downstream.

A similar vegetational succession has taken place here as around Santa Elena. The tropical forest has been almost entirely cleared, to be replaced by grasses and mechanised tonalmil.

The northernmost instance of floodplain patterning found to date in Northern Veracruz may be seen along the Estero la Llave, ca. 30 kms south of Tampico (Figure 3:7). The levee is low on the left bank and a secure zone along the top is hardly apparent. The raked pattern of canals down the backslope is the most obtrusive feature. One would have little difficulty naming the intervening spaces drained fields.

## An Hydroagricultural System

The preceding instances and others that might be described allow us to deduce a common pattern and hypothesise an intensification strategy in the construction and management of such a system. The pattern is summarised in Figure 3:8. The typical slope from the top of the levee to the backswamp is divided into four segements. The first (a) is the portion that is not flooded except in the sessional year where highwater is several metres above the mean, such as that indicated for the Nautla River in 1955 (Table 3:1). It varies in width, which is probably proportional to its height. This, in turn, would be a function of the flood-level maxima, but also the way water moves out of the main stream during an extraordinary flood. The water would tend to move more readily over an outside bank on a meander and hence

Figure 3:6 Detail out of the floodplain of the Estero, as marked on Figure 3:5

Figure 3:7 Patterning in the floodplain of the Estero la Llave

deposit more sediment there, leaving a higher levee. In most years the levee top can provide fairly secure sites for buildings and excellent land for subsistence crops, gardens and plantations of perennials.

The backslopes of many levees in Northern Veracruz are typically streaked with the remains of downslope-oriented canals and intervening fields, open on the upper end (b in Figure 3:8). This is similar to the raked wetland/*terra firma* margin seen frequently around complexes of Prehispanic canals and planting platforms in Central Veracruz. Some distance downslope the pattern is usually transected, forming the roughly rectilinear fields most often associated with ancient wetland agriculture (c in Figure 3:8).

There is a considerable regularity in the orientation of the canals near the Santa Elena ruins on the Nautla River. Most are laid out in two perpendicular directions, which diverge from our cardinal directions between 10° and 20° anticlockwise (Schmidt, 1977, p.52). Such a regularity is rare elsewhere in Northern Veracruz; an accordance between orientations of separate complexes, i.e. the imprint of a landscape organising principle, as was found in the wetlands of Central Veracruz (Siemens 1983a), is not yet apparent.

Collector drains, often straighter than one would expect from a natural water course, may lead into backswamps that were left unmodified (d in Figure 3:8). These in turn are drained by intermittent water courses that often run parallel to the main stream for considerable distances before they find openings through the levees. On the floodplains of meandering streams elements b. and c. may be occluded, extending right across the lowest portion of the floodplain. One of the rectilinear canals has usually been enlarged into a collector.

A network of canals such as those once used in many of the wetlands of coastal Veracruz would achieve rapid, sensitive results by rudimentary means. In a U.S. Department of Agriculture manual on drainage, which may perhaps be taken as representative of recent practical wisdom in this respect (U.S.D.A 1973), open drains are considered preferable to underground ducts for the drainage of large flat areas with gentle gradients. The configurations recommended are similar to those of the vestigial networks in northern Veracruz: a general parallelism in the channels, a herring-bone pattern across convex slopes, such as in an occluded floodplain, and dendritic hook-ups to main collectors. With a falling level in an adjacent main stream late in a wet season, each of the canals in such a network deflects the groundwater around it downward and efficiently conducts it downslope. It is thus brought down faster than it would descend without canalisation, allowing an earlier beginning in cultivation on the intervening fields than

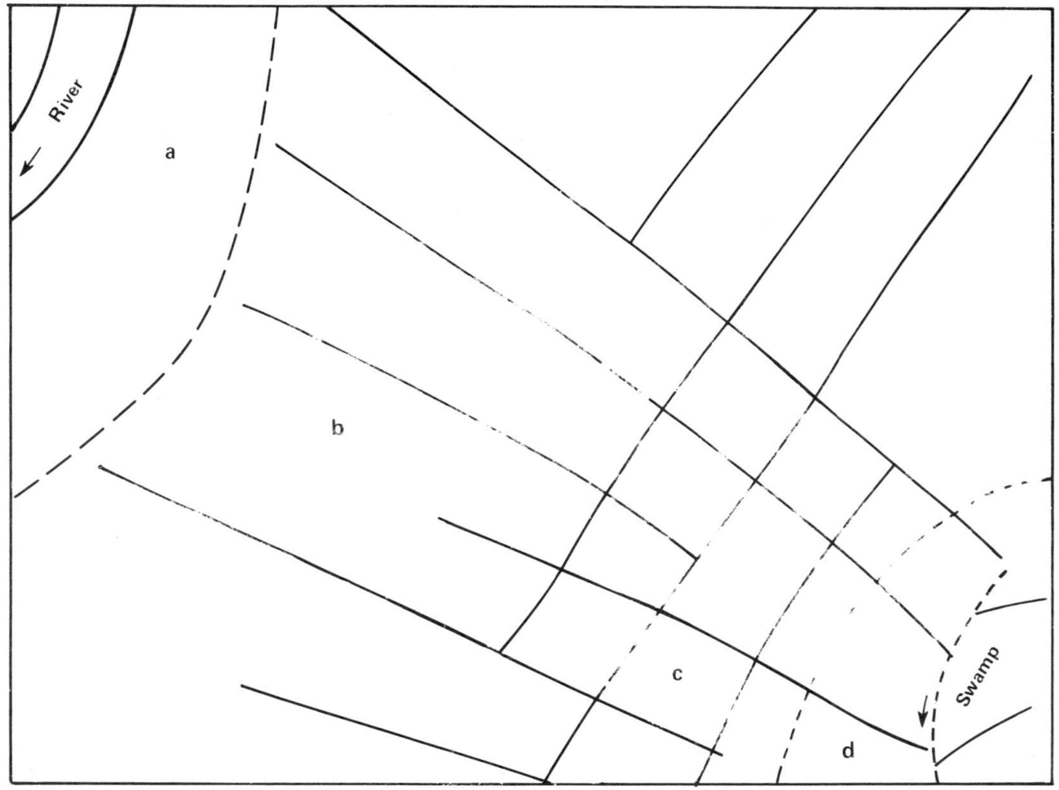


Figure 3:8 Schematic diagram of the hydroagricultural pattern hypothesised for levee backslopes in Northern Veracruz

would otherwise be feasible - the main point of the exercise, in modern as in ancient times.

An additional hydrological function is likely for the ancient system. The drainage function could be impeded at an advantageous topographic location and at a time during the dry season when the water level had fallen to a critical low point, by simple dams across the main collector drain. The descent of the water table could thus be retarded and kept within range of the roots of growing crops. Canoes could continue to circulate and farmers could practise scoop irrigation and fertilisation, which are common expedients in surviving Mesoamerican *chinampa* agriculture.

An annual floodplain dry season cropping cycle might therefore be thought of as beginning on the levee tops and ending on the lowest fields in the backswamp. There is as yet no direct evidence regarding the actual crops involved in the wetland agriculture of Northern Veracruz; investigations in other wetlands of the Mesoamerican Lowlands have confirmed maize and cotton (Puleston 1977), but beyond that there are only a series of possibilities, including root and tree crops (Wiseman 1983). The levee would be accessible throughout a normal year and could be expected to yield well in the wet season as well as the dry. It might be devoted to perennials as it often is now. With the end of the wet season agriculture would begin a migration downslope. Canalisation would speed access and transplantation would minimise the time an individual crop would need to be in the field. As the dry season progressed and the water level dropped further the lowest sections of the system would become accessible, the areas closely cross hatched by rectilinear remains. It is here that impedance of the main collector drains would have its greatest effect. By the time the crops were ready to be harvested the next wet season might well be underway again and the water level rising.

Such a system may be seen as an enhancement of the high-risk dry season agriculture without canalisation evident on floodplains in various lowland Mesoamerican locations. Coe and Diehl, for instance, have described in some detail the dry season agriculture that occurs in the floodplains of the Coatzacoalcos River (1980, pp.69, 147). The levee backslopes are planted between November and February with a crop known in the region as *tapachol*; harvest comes in May or June. Some low-lying areas are planted even later, in March, with a particularly risky crop known as *chamil*. In the Coatzacoalcos floodplain there is a great deal of land suitable for dry season cropping of this sort - without canalisation. The environs of the famous San Lorenzo site could have supported, the authors maintain, many more people than it did during Olmec times, in fact they see in this agriculture, "the most important clue to the rise of Olmec civilisation at San Lorenzo Tenochtitlan." (Coe and Diehl 1980, p.43). It is small wonder then that air surveys and vertical air photo analyses have so far revealed few signs of patterning in the floodplain of the Coatzacoalcos or similarly endowed floodplains along the other large rivers flowing into the southern portion of the Gulf of Mexico. The intensification represented by the construction and maintenance of a system of canals and planting platforms was not necessary.

An instance of dry season agriculture on the margin of wetland against terra firma has recently been investigated by this author near the village of El Palmar in Central Veracruz (Siemens 1983c). This activity is complementary, almost incidental to wet season agriculture on neighbouring terra firma, the food-producing mainstay of the community. It is subject to the vagaries of the region's hydrological regime and numerous other risks - more in all, it seems, than wet season agriculture. Yet given sufficient incentive the risks could be reduced by canalisation and even more by the construction of a complex of planting platforms. The remains of just such a complex are clearly visible in the middle of the wetland.

Some interesting observations on dry season agriculture have been made among the Kekchi Maya in Belize by Wilk (1981). It is known as the hunger-killer and is carried out on those portions of levee slopes judged neither too wet nor too dry. This is a gamble: a short dry season will mean the crop sits flooded out in a field too far downslope, a long dry season will mean desiccation in a field too far in the other direction. The best strategy seems to be to spread the planting as much as possible along the slope in order to ensure at least some yield. Planting begins some time in November, before the rains are over, and continues in stages until the rains appear to be diminishing. The harvesting is commensurately staggered. The yields per hectare are usually lower and more variable than those of wet season agriculture - a reflection of the high risks - but their timing is advantageous. They may constitute a critical supplement to the yields of a meagre previous wet season crop or at least allow the household to leave stored reserves untouched. In any case, there is little to occupy the farmer during the dry season. Here again, given a sufficient impetus canalisation and subsequent measures could be undertaken to secure this agriculture.

Considering these analogues helps one to envisage the preconditions for a move into wetland. They point towards the intensification process that has been hypothesised and underline its critical role. By means of it the need for elaborate "hedging of bets" could be reduced; more food could be produced more predictably.

As already indicated, the function of a risk-reducing system of canals and fields on a levee backslope may be designated as both drainage and irrigation, but only in special senses. It does not seem to have involved dyking and anything like a permanent lowering of the water table such as in land reclamation carried out in North Central Europe or Britain from medieval times to the present. The pumping technology was lacking. It may become apparent on further investigation that where wetlands are located not far from hill land, spring water or surface run-off was channelled into the backswamps to keep their water levels up, somewhat like what has been done traditionally along the southern perimeter of the *chinampas* on former Lakes Xochimilco and Chalco. It would seem more likely, however, that such water would be exploited in the hills by means of cross-channel terracing and irrigation of the kind Wilkerson has described west of Santa Luisa (Wilkerson 1979) or conducted to the lower slopes of *terra firma*, as may be inferred from the evidence for irrigation in lowland Central Veracruz cited by Sanders (Sanders 1953). The vestigial canal networks discussed here are perhaps best thought of as representing not so much an obliteration of the swamp or what has been commonly termed land reclamation, as a manipulation and an enchancement of the swamp regimen itself.

One may envisage considerable short-term resilience in such a system. By varying the time of closure of the collector drain during the dry season the cultivators could ameliorate the effect of yearly variations in precipitation within the catchment area. The normal annual flooding need not damage the system extensively. Even the effects of an extraordinary flood would be outweighed to a considerable extent by the benefits of deposition. With such natural renewal and additional muck scooped out of the canals, fallowing would be unnecesssary. It would be necessary, however, to expend considerable labour every year to clean and repair canals as well as to keep the planting surface at desired levels. Longer-term variations could also be adjusted to: by gradual deepening of the canals one might counteract a decrease in average precipitation or a drop in sea level; by gradual building up of the surfaces of the fields the opposite could be achieved.

The height of field surface build-up that may be imputed to the field remains visible from the air, in Northern Veracruz as elsewhere in the lowlands, is a critical question. It makes a good deal of difference to the length of time the fields may be assumed to have been accessible each year, how one thinks of the way they were managed and what overall significance they can be given within ancient subsistence systems. Excavations will need to be carried out, of course, but from the considerable water level fluctuations indicated, the gentle topographic slopes apparent from many levee-tops to their adjacent backswamps, the absence of evidence for the dyking that would be necessary under the circumstances to control water levels fully all year round, as well as the sheer mass of material that would have to be moved to build up the surface of the large fields encountered any appreciable amount, all this leads one to postulate a seasonal use, annual flooding and drying out.

A comparison with the context of the field complexes in the Maya region comes to mind at this point. Water level fluctuations in the karstic floodplains on the western and eastern flanks of the Yucatan Peninsula are somewhat less than those reported for the Northern Veracruz floodplains examined on the ground, yet a similar set of factors as outlined above have for some time led this author to entertain the probability of the seasonal use of at least some of the Maya wetland fields (Siemens 1982). Some recent excavations in Northern Belize point in that direction (Lambert and Arnason 1979; Chapter 6; Bloom *et al.* 1983). Seasonal use is certainly plausible for at least the margins of the field complexes investigated in Central Veracruz (Siemens 1983c). Some of these complexes are under intensive study now and should yield new evidence in this respect. A comparison with the closely controlled and certainly built up *chinampas* of Central Mexico must be added. The water level fluctuations in the Maya area as in Veracruz are considerably higher than the *chinamperos* would feel happy with. They cultivate their fields year round; fluctuations are kept well below a metre.

Flexibility seems to be the key to the conceptualisation of the possibilities in this regard as in so many other aspects of the interpretation of the evidence out of the wetlands. Seasonal use could mean various lengths of time for different segments of the wetland. With transplantation, use of the right strains and careful timing and location a succession of staple crops could be produced during one dry season in any given wetland. Moreover, with a progressive build-up the yearly period of access to any part or the whole of a wetland could be increased - under certain topographic and hydrological circumstances it could be extended the year round. Conversely, social stress, a decrease in requirements or other factors could lead to neglect, gradual erosion of the surfaces of the field, and a progressively shorter period of access to the system. The cultivators could withdraw entirely from the canalised and built-up portions of the wetlands, as actually happened, obviously and restricted themselves to such use of the unsecured margins as was necessary. The untended canals and platforms would soon degrade to an extent where reactivation would be very laborious.

The evidence for wetland agriculture in coastal Northern Veracruz need not suggest anything like Wittfogel's *hydraulic agriculture*, but, rather, something akin to his *hydroagriculture*, i.e. local facilities, built by families or groups of neighbours, linked perhaps into contiguous networks, certainly increasing the food supply but not dependent on elaborate social control (Wittfogel 1957). Nevertheless, the complexes do vary in their configuration. That north of Santa Elena is more strictly rectilinear and consistently oriented than others in Northern Veracruz. Some differences in central influences, some variations in organising principles or at least some differences in chronology are suggested.

The morphology of Northern Veracruz raised field complexes also suggest an intriguing trans-lowland relationship. The many complexes on the floodplains of the karstic rivers that emerge on the two sides of the Yucatan Peninsula are also mostly without very much rectilinear regularity or consistent orientation. One exception, across the Candelaria River from the remains of the capital of Acalan, Itzamkanac, is aligned to cardinal directions some 8$^{\circ}$ anticlockwise of our own (Siemens and Puleston 1972). There is nothing like the consistency in orientation of field vestiges (ca. 15$^{\circ}$ clockwise) within and between the wetlands of Central Veracruz (Siemens 1983a). In general terms, the fields in the floodplains of the Maya Lowlands and those of Northern Veracruz, certainly those around Tamiahua, have a rudimentary appearance as far as engineering is concerned. These similarities correspond to links between the Mayas and the Huastecas postulated some time ago by Ekholm on the basis of ceramics pertaining to Mamom and Chicanel phases (Ekholm 1944), i.e. between 1000 B.C. and A.D. 400, which is also known as the Formative Period in Mesoamerican chronology (Sanders and Price 1968). It may well be, therefore, that the eventual archaeological investigations of raised field stratigraphy in Northern Veracruz will come on Formative dates as similar investigations have done in several locations within the Maya Lowlands (Siemens and Puleston 1972; Puleston 1977; Turner and Harrison 1983).

The emphasis throughout this discussion has been strictly on the agricultural features and practises of the wetlands. The full range of subsistence activities of which they were a part have been elaborated in another context (Siemens 1983b). Wet season cropping on hill lands, together with its own <u>tonalmil</u>, would be going on alongside the cultivation of the wetland. Irrigation and terracing would probably have served to intensify such cultivation in various locations. Around the households, whether located on <u>terra firma</u> or on levees, small plots must have been in cultivation. In addition, the fish trapped in the canals with the recession of floodwaters could be harvested, water fowl could be hunted and other easily imaginable ancillary activities undertaken. It would seem that enhancement of the variety and fit of subsistence activities, as much as the increase of total output would be the objective of intensification.

The early colonial chroniclers heard their Aztec informants describe the rich agricultural capacity of their coastal homeland in terms of well-orchestrated cropping (Duran 1964). Legend or otherwise, the description sharpens the appreciation of the actual evidence for such activity. Northern Veracruz had been a region of abundance. There had been plenty of water; the environment had offered food resources of many kinds. Most importantly,at any given time there had been some maize ripe and ready to pick, some with new ears, some just growing up and some newly sprouted. In such a place there could never be famine.

## Acknowledgements

The research on which this paper is based was supported by the Social Sciences Research Council of Canada and the Instituto Nacional de Investigaciones sobre Recursos Bioticos in Jalapa (INIREB), Veracruz. The author is now engaged in a cooperative investigation of the prehistoric use of wetlands in Veracruz with colleagues at INIREB.

## Bibliography

BLOOM, P.R., M.POHL, C. BUTTLEMAN, F. WISEMAN, A. COVICH, C. MIKSICEK, J. BALL and J. STEIN 1983, Prehistoric Maya wetland agriculture and the alluvial soils near San Antonio, Rio Hondo, Belize, Nature 301, 5899, 417-419.

CAMOU, H.E. 1981, Los Sistemas de Produccion de Maiz en el Norte de Veracruz, Centro de Ecodesarrollo, Mexico, 47pp.

CARTA GEOLOGICA de MEXICO, 1977.

COE, M.D. and R.A. DIEHL, 1980, Peoples of the Rivers, University of Texas Press, Austin, 37-39.

COMPANIA MEXICANA de AEEROFOTO, 1954, Flight 1130 (1:35,000)

COMPANIA MEXICANA de AEROFOTO, 1968, Flight 1815 (1:10,000).

DETENAL, 1976, Vertical air photos, Zone ZAR-351, Flight line 5, 1:90,000 (Feb. 14).

DIAZ, B. 1976, The Conquest of New Spain, Penguin Books, Harmondsworth, 107-108.

DURÁN, F.D. 1964, The Aztecs, Orion Press, New York, 137-138.

EKHOLM, G.F. 1944, Excavations at Tampico and Panuco in the Huasteca, Mexico, The American Museum of Natural History, Anthropological Papers, XXXVIII, 87-88.

GARCIA, E. 1970, Los climas del estado de Veracruz, Anales del Instituto Biologico, Universidad National Autonoma, 41, Seria Botanica, 3-42.

von HUMBOLDT, A. 1811, Political Essay on the Kingdom of New Spain, J. Black (translator), Longman, London, 250-254.

LAMBERT, J.D.H. and J.T. ARNASON 1979, Drained fields in the New River floodplain and their agricultural potential - past and presnt, Ms. on file, Department of Biology, Carelton University, Ontario.

PALERM, A. 1953, Etnografia antigua totonaca en el Oriente de Mexico, In, Huastecas, Totonacos y sus Vecinos, edited by I. Bernal and E.D. Hurtado, Revista Mexicana de Estudios Antropologicos XIII, 163-173.

PULESTON, D.E. 1977, The art and archaeology of hydraulic agriculture in the Maya Lowlands, In, Social Process in Maya Prehistory: Studies in Memory of Sir Eric Thompson, edited by N.Hammond, Academic Press, New York, 449-467.

RZEDOWSKI, J. 1978, Veetacion de Mexico, Limusa, Mexico, 154pp.

SANDERS, W.T. 1953, The Anthropogeography of Central Veracruz, In, Huastecas, Totonacas y sus Vecinos; Revista Mexicaa de Estudios Antropologicos XIII, 27-28.

SANDERS, W.T. and B. PRICE 1968, Messoamerica: The Evaluation of a Civilisation, Random House, New York, p.15.

SARTORIUS, C. 1961, Mexico about 1850, Brockhaus, Stuttgart, p.8.

SCHMIDT, P.J. 1977, Unsistema de culturo intensivo en la cuenca del Rio Nautla, Veracruz, Boletin de INAH 20, 50-60.

SCHOLES, F.C. and R.L. ROYS 1968, The Maya Chontal Indians of Acalan-Tixchel, University of Oklahoma Press, Norman, 123pp.

SECRETARIA de RECURSOS HIDRAULICOS, MEXICO 1970, Boletin Hidrologico 42, 11690.2.18.

SIEMENS, A.H. (editor) 1977, The Rio Hondo Project: An investigation of the Maya of Northern Belize, Journal of Belizian Affairs 5.

SIEMENS, A.H. 1978, Karst and the pre-Hispanic Maya in the Southern Lowlands, In, Pre-Hispanic Maya Agriculture, edited by P.D. Harrison and B.L. Turner II, University of New Mexico Press, Albuquerque, 414pp.

SIEMENS, A.H. 1982, Pre-Hispanic agricultural use of the wetlands of Northern Belize, In, Maya Subsistence: Studies in Memory of Dennis E. Puleston, edited by K.V. Flannery, Academic Press, New York, 205-225.

SIEMENS, A.H. 1982, Aprovechamiento agricola precolombino de tierras inundables en el norte de Veracruz, Biotica 3, 343-357.

SIEMENS, A.H. 1983a, Oriented raised fields in Central Veracruz, American Antiquity 48 (1), 85-102.

SIEMENS, A.H. 1983b, Wetland agriculture in the lowlands of pre-Hispanic Mezoamerica, Geographical Review 73 (2), 166-181.

SIEMENS, A.H. 1983c, Reducing the risk: seasonal cultivation of wetland margins, submitted for forthcoming volume edited by S. Gliessman.

SIEMENS, A.H. and D.E. PULESTON 1972, Ridged fields and associated features in Southern Campeche: new perspectives on the lowland Maya, American Antiquity 37, 228-239.

STRAESSER-PEAN, G. 1968, Ancient sources on the Huasteca, Handbook of Middle American Indians, Vol. II, Part 2, University of Texas Press, Austin, 588pp.

THOMPSON, W. 1946, Recollections of Mexico, Wiley and Putnam, New York, 12pp.

TURNER II B.L. 1974, Prehistoric intensive agriculture in the Mayan lowlands, Science 185, 118-124.

TURNER II B.L. and P.D. HARRISON 1983, Maya Raised-Field Agriculture and Settlement at Pulltrouser Swamp, Northern Belize, University of Texas Press, Austin (in press).

UNITED STATES DEPARTMENT of AGRICULTURE 1973, Drainage of Agricultural Land, Water Information Center, Port Washington, New York, 75-79.

WILK, R.R. 1981, Agriculture, Ecology and Domestic Organisation Among the Kekchi Maya, Ph.D. dissertation, University of Arizona, 203-236.

WILKERSON, S.J.K. 1979, Huastec presence and cultural chronology in North Central Veracruz, Mexico, Actes du XLII Congres Internacional des Americanistes 38, 31-47.

WILKERSON, S.J.K. 1980, Man's eighty centuries in Veracruz, National Geographic 158, 2, 203-231.

WISEMAN, F.M. 1983, In, Maya Raised-Field Agriculture and Settlement at Pulltrouser Swamp, Northern Belize, edited by B.L. Turner II and P.D. Harrison, University of Texas Press, Austin (in press).

Chapter 4

# SO GREEN AND LIKE A GARDEN: INTENSIVE AGRICULTURE IN ANCIENT VERACRUZ

S. Jeffrey K. Wilkerson

Abstract

The coast of the state of Veracruz, Mexico, is divided into four zones; the North Gulf Area, the North Central Gulf Area, the South Central Area and the South Gulf Area. Investigations as to why at the time of Contact Cortes and others reported lush green country whereas today the vegetation is dry and burned in the dry season, are made. Evidence suggests that the Totonac capital, Zempoala, in the North Central Gulf Area, with an estimated population of between 80,000 and 120,000 was green because it was based on year round intensive drained field agriculture. Exploration of drained field systems on the Tecolutla and Nautla Rivers, close to Zempoala, begin in 1968 and drained fields around Santa Luisa were documented in 1974. These fields are on the narrow alluvial plains of the lower Tecolutla River. Other fields were found at Santa Elena and in the Nautla drainage basin. They date between 600-1100 A.D. The fields and canals have the double function of providing water for dry season agriculture and removing excess water during the wet season. All of these fields are located on terraces near the mouths or on estuaries of tidal rivers. Hill terraces to the west of the fields were probably cultivated simultaneously. Both Santa Luisa and Santa Elena are within the sphere of influence of the large site of El Tajin and it is likely that food and tribute crops grown on these fields and terraces supported the urban population. Cotton and cacao are thought to have been important non-subsistence crops. The Santa Luisa drained field system is the largest and it may date to the Tecolutla (ca. 1-300 A.D.) or the Cacahuatel Phase (ca. 300-600 A.D.). Lacustrine fields have also been found near the site of Las Higeras in the same zone. More fields occur in marshes near the mouth of the Rio Tuxpan. In the North Gulf Area additional field and terrace systems have been observed from the air. Further south, in the South Gulf Area, on the Rio Coatzacoalcos and Rio Tonala, no fields have been documented which suggests that here there were no sufficiently large and centralised social units in existence at the time drained field technology was current.

Resumen

La costa del estado de Veracruz, México, se divide en cuatro zonas: el Area del Norte del Golfo, el Area Norte Central del Golfo, el Area Sur Central del Golfo y el Area Sur del Golfo. Se investiga por qué en el momento del Contacto Cortés y los demás informaron sobre un campo exhuberante y verde cuando hoy día la vegetación está seca y quemada en la estación seca. La evidencia sugiere que la capital Totonaca,Zempoala, en el Area Norte Central del Golfo con una población estimada de entre 80.000 y 120.000 fue verde debido a que estaba basada sobre una agricultura intensiva de campo drenado durante todo el año. La exploración de sistemas de campos drenados en los ríos Tecolutla y Nautla, cerca de Zempoala, comenzó en 1968 y los campos drenados alrededor de Santa Luisa fueron documentados en 1974. Estos campos estan sobre las llanuras aluviales del río Tecolutla inferior. Otros campos se encontraron en Santa Elena y en la cuenca del Nautla. Estan fechados entre 600-1100 D.C. Los campos y canales tienen la doble función de abastecer agua para la agricultura de estación seca y drenar el exceso de agua durante la estación húmeda. Todos estos campos se ubican sobre terrazas, cerca de las desembocaduras o sobre los estuarios de los ríos de marea. Las terrazas de ladera al oeste de los campos probablemente se cultivaron simultáneamente. Tanto Santa Luisa como Santa Elena estan dentro del círculo de influencia del gran sitio de El Tajín y es probable que los alimentos y los cultivos de ofrenda cultivados en estos campos y terrazas mantuvieran a la población urbana. Se piensa que el algodón y el cacao fueron cultivos importantes de no-subsistencia. El sistema de campo drenado de Santa Luisa es el más grande y puede remontarse hasta el Tecolutla (alrededor de 1300 D.C.) o a la Fase Cacahuatel (alrededor de 300-600 D.C.). También se han encontrado campos lacustres cerca del sitio de las Higueras en la misma zona. Existen más campos en los pantanos cerca de la desembocadura del río Tuxpan. En el Area Norte del Golfo se han observado desde el aire más sistemas de campos y de terrazas. Más al sur, en el Area Sur del Golfo, en el río Coatzacoalcos y el río Tonalá no se han documentado campos, lo que sugiere que en este lugar no existieron unidades sociales suficientemente grandes y centralizadas en la misma época de la tecnología de campo drenado.

In the dry spring of 1519 Hernan Cortés landed on the coast of Veracruz, not far from today's port of the same name. His first march inland through the brown bush of the dry season brought him to the Totonac capital of Zempoala.

Surprise at its size, and greeness was general in the Spanish army. The astonished soldier Bernal Díaz (1963:107) stated:

> ...as we came among the houses we saw how large a town, larger than any we had yet seen, and we were full of admiration. It was so green with vegetation that it looked like a garden; and its streets were so full of men and women who had come out to see us that we gave thanks to God for the discovery of such a country.

Since colonial times it has been customary to divide the coast of Veracruz into two regions: Barlovento (windward) and Sotovento (leeward). Initially a reflection of the concerns of sailing masters, it also denotes geographic divisions of the coast (Figure 4:1). The dividing point, just north of Zempoala, is the abrupt Sierra de Chiconquiaco.

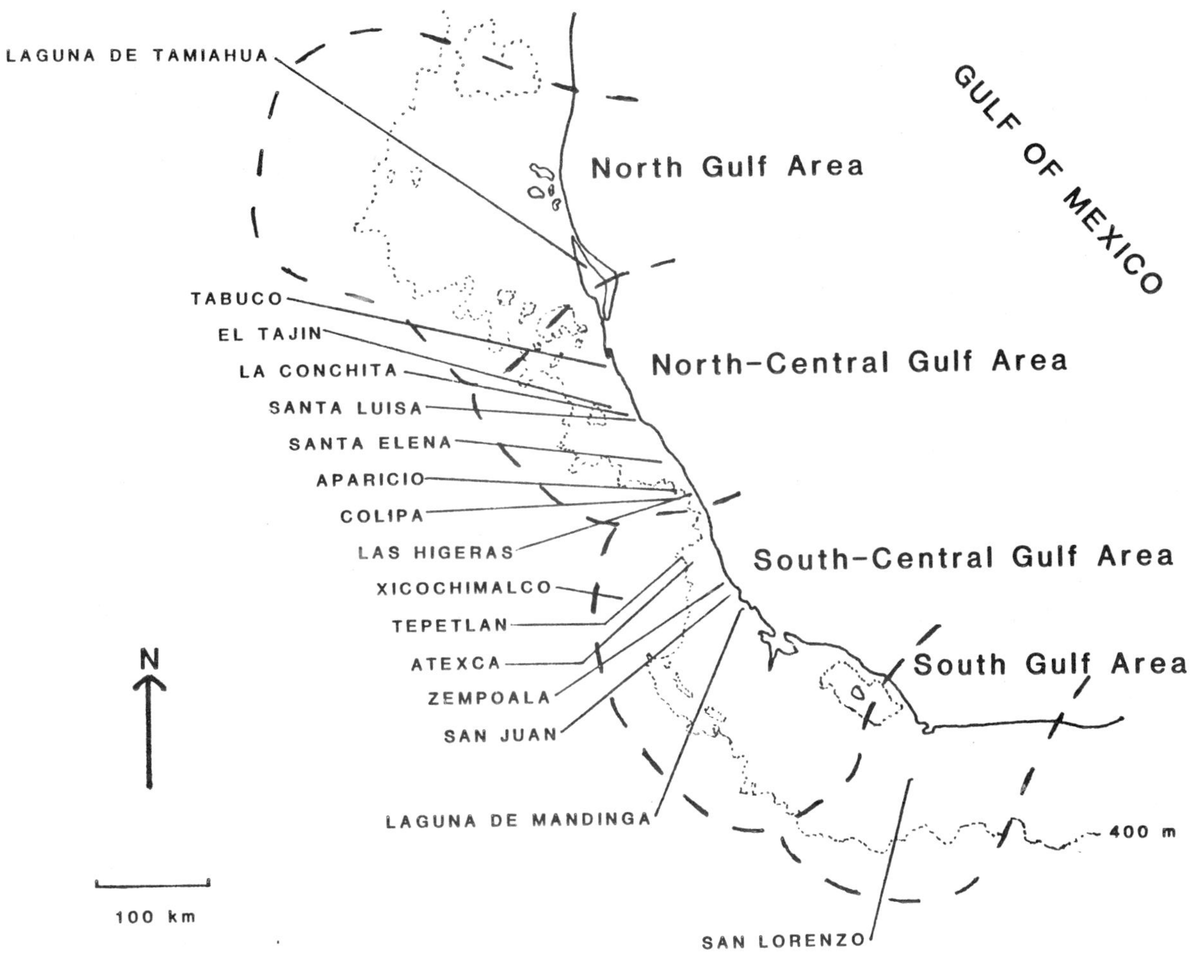


Figure 4:1 Pre-Columbian areas and sites along the Gulf Coast of Mexico

Thousand foot peaks of this eastern terminus of Mexico's transverse volcanic chain literally abut the Gulf at Quiahuitzlan, near where Cortés landed. The Sierra is an effective barrier to the predominantly northerly winds which dominate from September to February. Annual precipitation reaches 4500 mm on the northern flanks of the mountains while portions of the rain shadow area to the south register precipitation as low as 500 mm.

Zempoala is situated on the coastal side of the rain shadow area and to encounter the original inhospitable thorn forests one only has to walk from the ancient city's centre to the margin of the modern irrigated sugar-cane fields, or a few paces back from the nearby Actopan River. At the time of Cortés' arrival, in the dry season, the natural vegetation would have been burnt brown rather than green or garden-like. How then did this "delightful paradise" with year-round abundance, described by Torquemada (1620:I:396), come to be here?

Modest populations subsisted on extensive riverine resources near Zempoala from at least Early Formative times (García Payón 1966; Wilkerson 1981). But, however old its roots, a city whose population is estimated at between 80,000 and 120,000 in size must have had a large and productive subsistence base. Zempoala most certainly had a system of intensive agriculture as well as the political clout to draw on a broad supporting region for tribute goods. In turn it paid tribute to the Aztec state from the time of its conquest by Moctezuma I in the mid-fifteenth century (Kelly and Palerm 1952:269-272).

Evidence suggests that Zempoala was based on intensive agriculture in a region of low rainfall. This was made possible, as it is today, by diverting water onto the alluvial soils of the area. Success at this endeavour would have led to the increased size and political importance of Postclassic Zempoala. Although there are references to a water distribution system in the houses of the elite (Cortés 1979:2; Torquemada 1629:I:396), and exploration as far back as the end of the last century (see Strebel 1884; Galindo 1912), there has not been a major study of Zempoala's irrigation system(s), but García Payón found ancient canals which were used until government sponsored redevelopment occurred (personal communication 1965). Beneath the sugar-cane fields, the modern canals, and the twentieth century town lie the remains of what may have been the most important intensive agricultural system on the Veracruz coast when Europeans arrived.

The origins of such elaborate, and labour intensive technology are much older than the Postclassic Period (ca. A.D. 900-1520) in the ecologically diverse and fertile coastal plains of Veracruz. Let us examine some of the Classic Period (ca. A.D. 300-900) antecedents and look at occurrences elsewhere.

## Field Systems on the Tecolutla and Nautla Rivers

Beginning in 1968, explorations on the lower Rio Tecolutla (Wilkerson 1972) have yielded information on a series of important sites and intensive agricultural practices of the Classic Period. A series of field systems centred around Santa Luisa (Figure 4:1) were first encountered in 1974. The setting of Santa Luisa is typical of the environment where one would expect to find large field systems. Located a few kilometres from the Gulf coast, on the narrow alluvial plains of the lower Tecolutla River, the site stretches along and away from the river's course. Although today the north bank is a gently sloping terrace ending in a brackish marsh, it was once a series of deltaic islands near the present coastline through which the river flowed to a sheltered aggrading embayment. A number of these islands were inhabited from the mid-Archaic Period (Wilkerson 1973, 1975). Sometime between the Early Formative and the Late Formative the river swung eastward towards its present course and this, combined with a rising sea level and a lack of currents in the embayment, caused it to silt in, expanding the dry areas north of the site.

During the Classic (ca. A.D. 300-900) and the local Epi-Classic (ca. A.D. 900-1100) systems of canals and fields were prepared on this newly created northern terrace of the river. Smaller, but nonetheless significant systems were also developed on the south side of the river and on the narrower floodplains of tributaries of the Tecolutla. Additionally terraces were constructed in the sandstone hills west of the site. It was a time of considerable planning, construction, and population increase.

Analysis of the major field systems at Santa Luisa (Wilkerson 1975-79), included studies of infrared photography (Figure 4:2). The results indicated some important features which may contrast with known lowland systems in the Maya region. Unlike many of these examples irrigated field systems of the Isla A & B phases (A.D. 600-1100) at Santa Luisa and Santa Elena (Schmidt 1977) in the Nautla drainage basin fifty kilometres to the south (Figures 4:3, 4:4 and 4:5 and Table 4:1) are: 1) located on terraces near the mouths and on estuaries of tidal rivers, 2) receive their freshwater supply from streams entering the deltaic area (Figures 4:6 and 4:7), and 3) may form a portion of a large subsistence-tributary system associated with a state society (El Tajín). These two Veracruz systems appear to have the double function of providing the necessary water for dry season agriculture (*tonamil*) and removing some excessive water, from around the raised cultivation surfaces during the wet season planting (*temporal*).

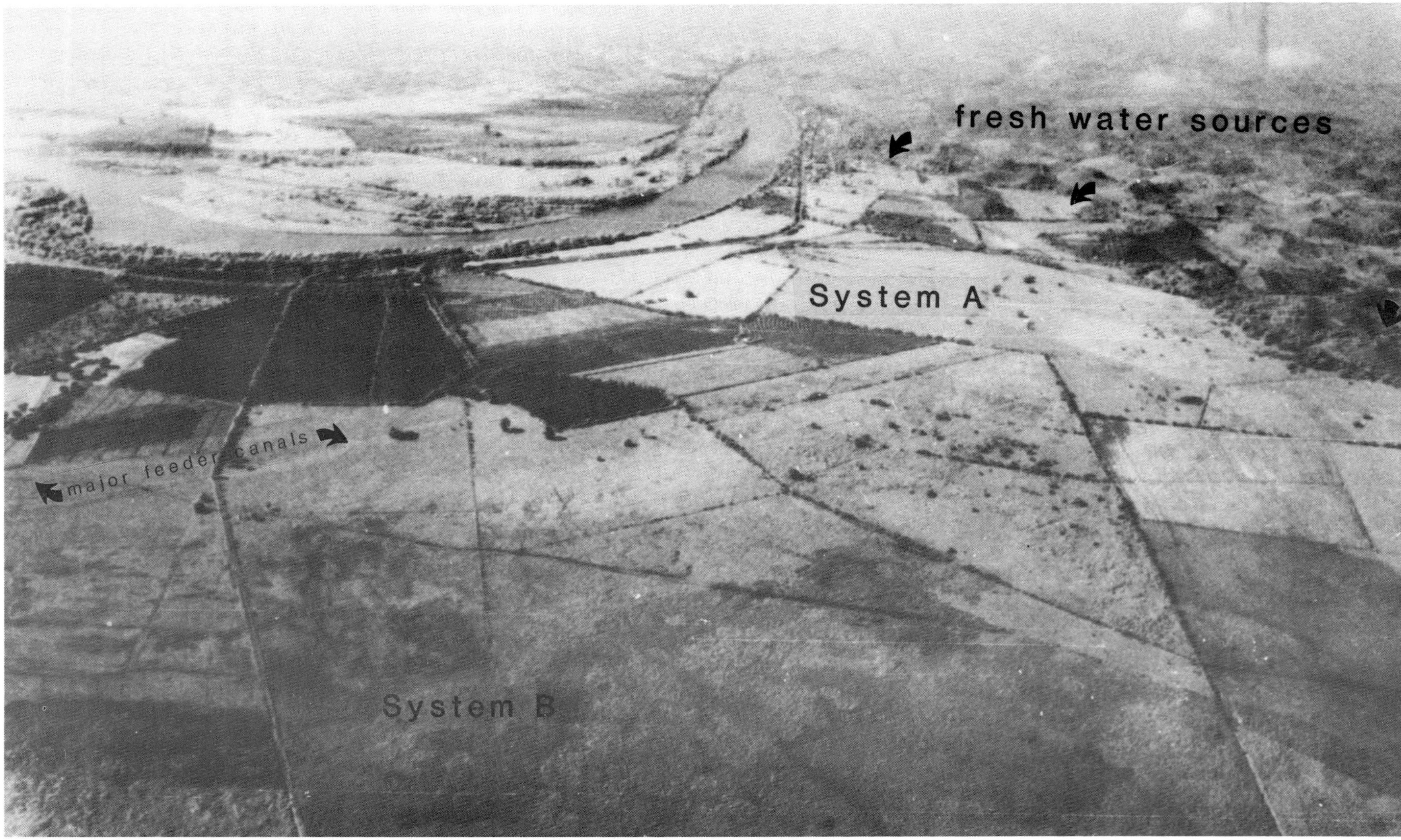


Figure 4:2 Low altitude infrared view of System B (foreground) and System A (right centre) at Santa Luisa

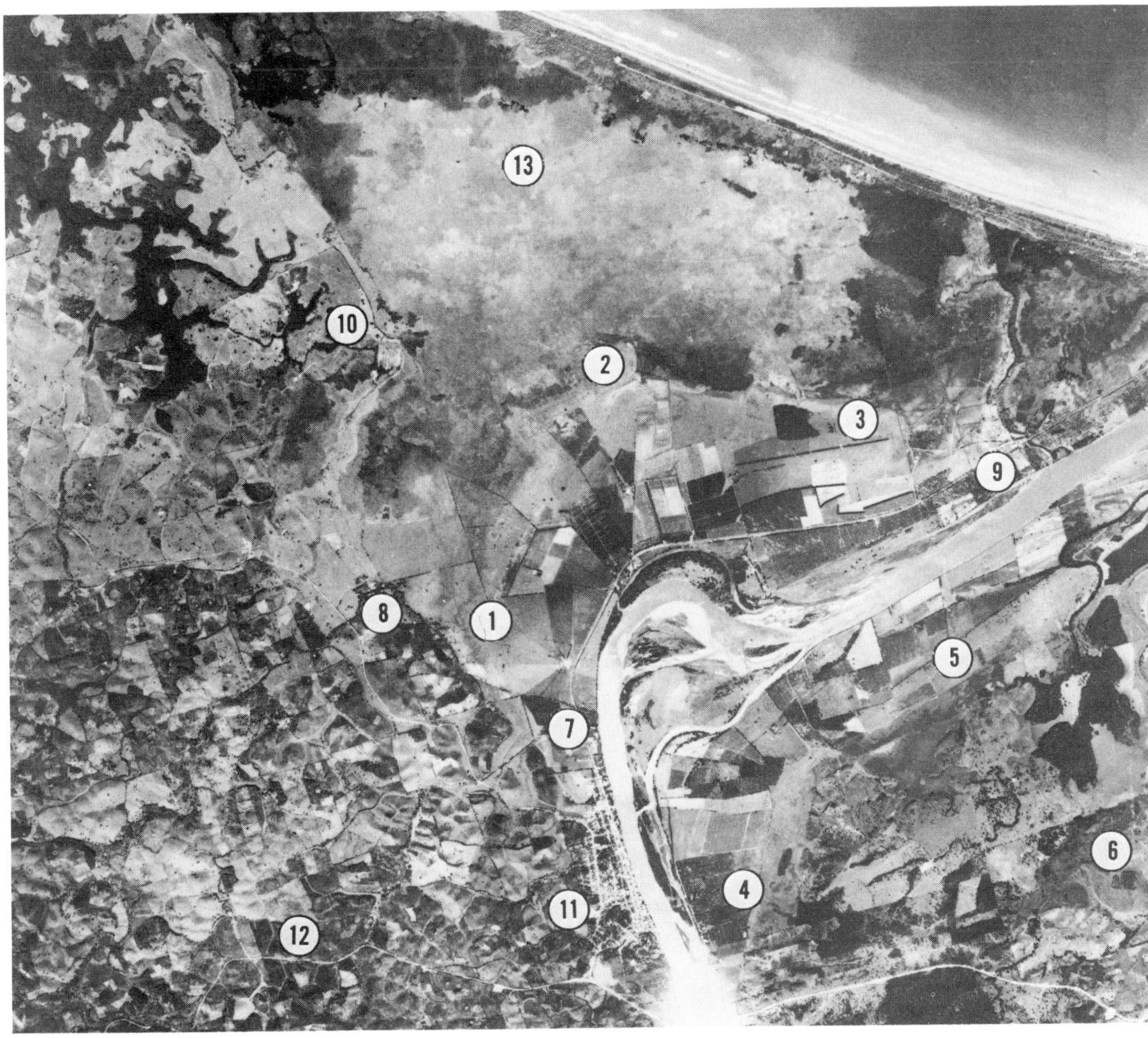


Figure 4:3 Aerial view from 6,600 feet altitude of the Lower Tecolutla drainage. 1) Santa Luisa System A; 2) Santa Luisa System B; 3) Santa Luisa System C; 4) Cacahuatal satellite site and sytem; 5) Santa Rosa satellite site and system; 6) El Mirador satellite site and system; 7) the architectural focus of Santa Luisa, living area stretches south beneath Gutierrez Zamora, northeast to the limit of hills, and eastward along the north margin of the Rio Tecolutla at least as far as 9); 10) the satellite site of La Isla; 11) modern town of Gutierrez Zamora; 12) area of excavated catchment terraces; and 13) swamp and former embayment

Figure 4:4 A portion of System C at Santa Luisa. Note canal patterns

Figure 4:5 The Santa Elena system on the Estero Tres Bocas just north of the Nautla River

| Type | Period | Phase range | Date |
|---|---|---|---|
| A | Postclassic | El Cristo-Cabezas | A.D. 1100-1520 |
| A-1 | Proto-Classic | Tecolutla | A.D. 1-300 to |
| | Early and Late Classic | Cacahuatal-Isla A and B | A.D. 1100 |
| B | Late Classic | Isla A & B | A.D. 600-1100 |
| C | Late Classic | Isla A & B | A.D. 600-1100 |
| D | Late Classic | Isla A & B | A.D. 600-1100 |
| G | Late Postclassic | Cabezas | A.D. 1300-1520 |

Table 4:1 Dating of field and terrace systems in North-Central Veracruz

Unlike some irrigation systems elsewhere, there are only a few discernible trunk canals or large-scale symmetrical alignments. Rather, there are rectangular blocks of canals frequently set at right angles to one another. Blocks near the freshwater source in the Santa Luisa system have an internal checkerboard alignment (System A) while those downslope (Systems B & C) are subdivided into parallel arrangements (Figure 4:8). One of the downslope blocks encompasses 3.9 hectares (Figure 4:4).

Water movement through the shallow canals, even allowing for the water level one metre lower in the Classic Period, as suggested by initial excavation in the system, would have been slow and siltation would have been a potential problem. Major floods, on approximately a fifteen year cycle (5 to 27 year parameters), also must have created the need for maintenance, labour and centralized direction. Occasional house mounds within the systems may represent caretaker locations for permanent canal maintenance.

Although the data are very preliminary it is possible to *tentatively* test some models of food production and population. The Santa Luisa field system, north of the Rio Tecolutla, encompasses approximately 6.75 square kilometres. The settlement associated with the canals is approximately 5.45 square kilometres, excluding the smaller satellite site of La Isla (approximately one square kilometre in size) which borders the irrigated area on the north. For purposes of this discussion the smaller settlements and canal systems on the south side of the river are not included as information on their extent and nature is incomplete. Preliminary data, however, suggests a similar ratio of occupied area to fields.

The region of hill terraces to the west was in use simultaneously with the irrigation system. The extent of these terraces is still unknown but some estimate of their probable area would have to enter into any final calculation of total food production potential. Certainly this potential would be less per unit area than that of the irrigation system. Given the poorer quality of the

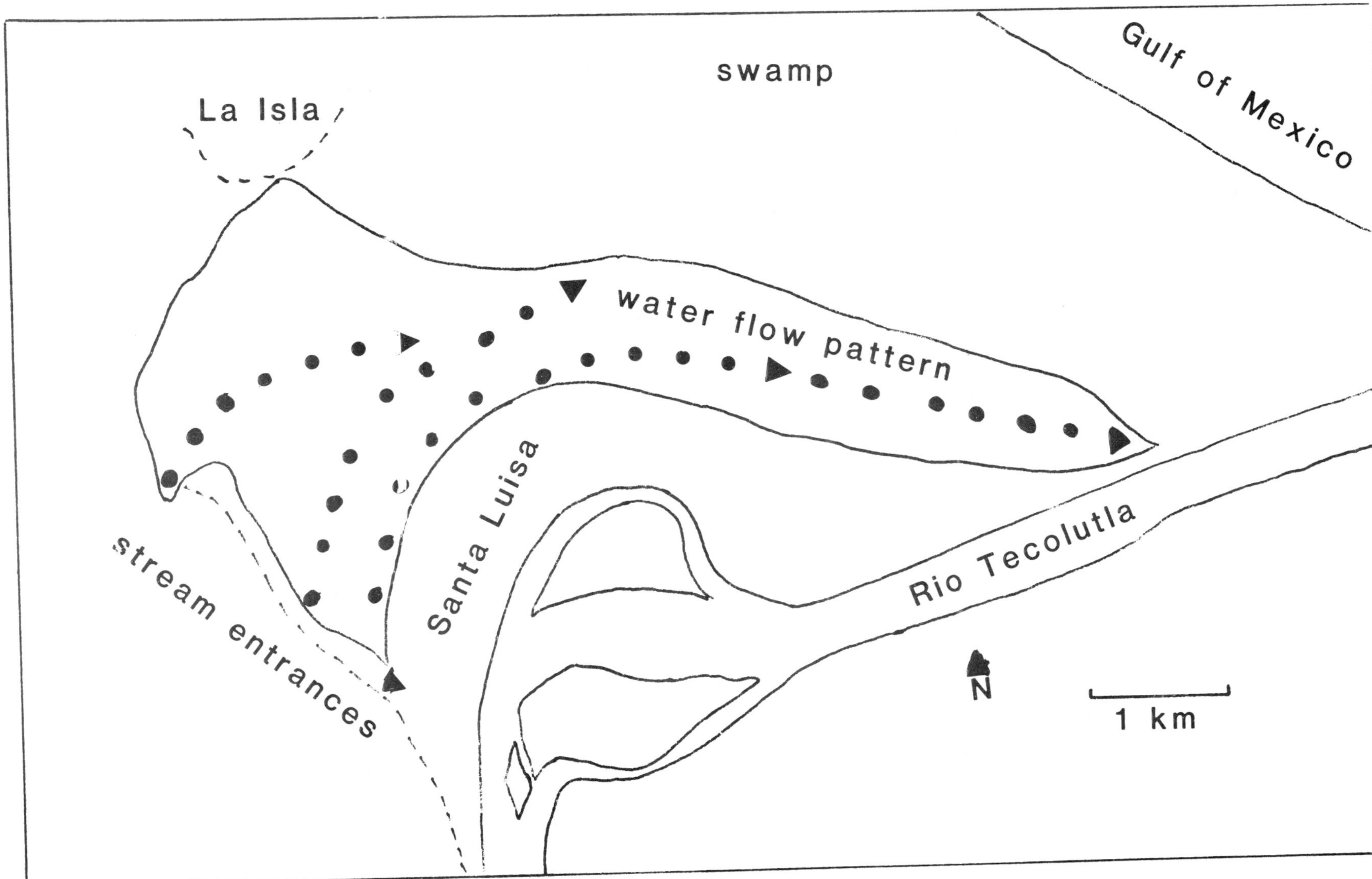


Figure 4:6 Water flow pattern in the Santa Luisa system

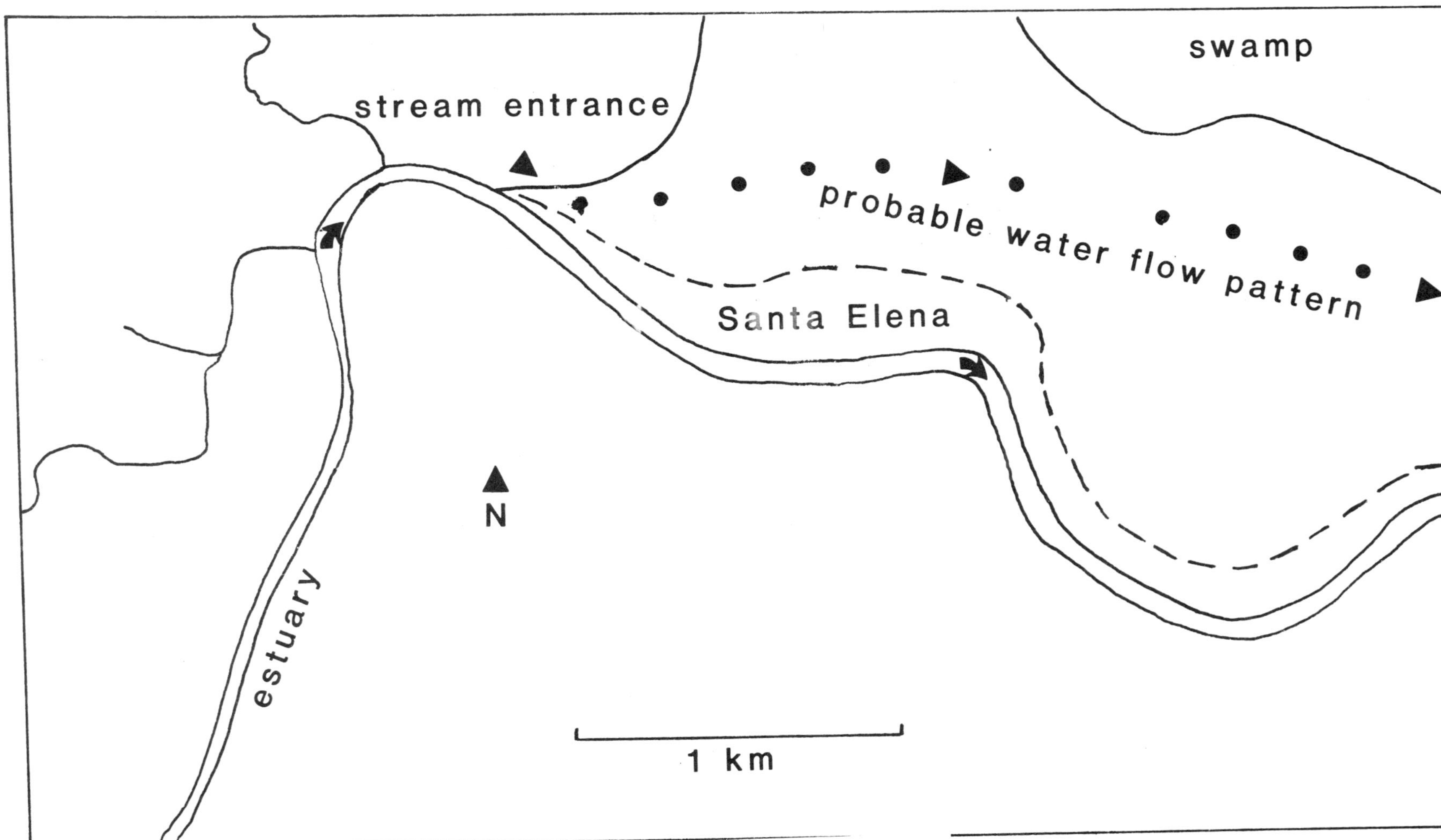


Figure 4:7 Water flow pattern in the Santa Elena system

Figure 4:8 A portion of System A at Santa Luisa

thin soils on the sandstone hills corn production is likely to have been between one and one-and-a-half metric tonnes per hectare per year. Yield data are based on <u>en grano</u> or shelled corn of the local 'criolla' variety grown in the municipalities of Papantla, Gutiérrez Zamora and Tecolutla by Totonacs in the years 1969-1975. The more fertile alluvial terrace soils of the irrigation system could produce about three tons per hectare per year. This would be based on a two ton yield for the spring dry season and a one ton yield for the summer wet season. Using these tentative figures the potential annual harvest of the irrigated area, assuming that corn was planted in the total area in both seasons, is 2025 metric tonnes.

It was proposed (Wilkerson 1975) that the canal blocks pertain to kin units, quite probably to extended families of patrilineages. Due to the deteriorated condition of the system and limited exploration time it has not been possible to confirm the uniformity of the blocks. However, initial field reconnaissance suggests that at Santa Luisa they are all of similar dimensions to the one which measured 3.9 hectares (at Santa Elena each unit may be nearer 3.2 hectares). Using Palerm's study (1955) of a highland system at Tecomatepec in the Valley of Mexico, where each family was allowed 0.86 hectares 4.54 families per block could be accommodated. Using early 20th century census figures the indigenous family in the region averages 5.9 members. Application of this number produced 27 individuals per block. This is consistent with lowland ethnographic evidence for extended families and implies a total population of 4,632 individuals for the habitation areas (Santa Luisa, La Isla, and hills) around the system.

The total harvest of 2025 tons of corn would, for purposes of this preliminary model, be equal to 437.2 kg per person per year. The daily intake would be 1.2 kg per person. This is at least four times the average amount consumed today in the region. Even allowing a thirty percent loss to rodents while in storage, a normal loss in North-Central Veracruz, the remainder is still about three times greater than current consumption. While these calculations assume total corn production, they do not include the yield potential of the large terraced area, the south bank system, the implications of intercropping, or swidden yields on the poorer soils behind the river terraces.

Although these preliminary figures may change with additional information from field exploration, or the application of different interpretive models, it appears likely that the potential food production exceeded the needs of the local population. This is exactly the situation one would expect if these systems were designed to produce surpluses or trade/tribute produce for a larger social unit. The centralized nature of regional social organization in the Late Classic Period may account for this phenomenon.

Data from archaeological reconnaissance and excavations indicate a high population density and very probably a state social organization centred at the large site of El Tajín (Wilkerson 1976; n.d.). Both Santa Luisa and Santa Elena are clearly within the El Tajín sphere of influence and domination and may have provided part of their production as tribute or for exchange. This possibility requires careful scrutiny and must be ultimately tested against more comprehensive field data. Additional infrared and black and white photography is required of the relic systems in contrasting seasons of the year. Further ground testing and pollen analysis should also be undertaken.

The type of non-subsistence crops grown needs to be examined. At the end of the Postclassic cotton was the major tribute item paid by the province of Tuxpan to the Aztecs. However, it was paid in finished blankets and garments. How far back in time this custom stretched is not known. At present evidence for the Classic Period is minimal although abundant clay spindle whorls suggest that it was a common commodity.

Cacao was another important crop. Intensive cacao production is often associated with irrigation and during the Postclassic and Early Colonial Periods it was most closely connected with the province of Soconusco and the Pacific coast (Gibson 1964:348). However, its coveted nature led to widespread attempts at plantation development elsewhere, often despite less than ideal climatic conditions. Cacao is a tribute and trade item *par excellence* that would enhance the economic power of any province. Its successful propagation in the El Tajín region would soon have led to extension of the cultivation area. But what evidence is there for cacao at El Tajin and Santa Luisa?

At El Tajín, the metropolitan centre of the region, there is direct evidence in the form of a carved panel depicting a ritual about a cacao tree (Figure 4:9). Originally from the Temple of the Niches, the major temple in the older portion of the city, is located today in the Museo de Antropologia in Jalapa. There is no doubt that the tree is cacao and not a maguey as suggested by Kamper (1972) because the fruit is ellipticoid and occurs all over the branches and on the trunk. Maguey holds its flowering stalks upwards in horizontal projections at and near the top of the stem only.

The ceremony on the carving portrays an animal (1), perhaps a feline or dog, crouching on a stepped platform (2) beneath the hand of a person (3) with a death figure held to his head-dress by a tumpline (4). The platform is in front of a mature cacao tree (5) with abundant fruit protruding from the limbs and trunk. A person (6)ascends the steps holding a torch-like object in his hands. The accoutrements are a fan and staff, symbols frequently associated with merchants and the gods of commerce. An

Figure 4:9 A carved panel from the Temple of the Niches at El Tajin depicting a ritual around a cacao tree

earth monster (7) resides under the platform. One of the two personages represented may be the Tajín equivalent of the Maya God M, also known as "the black god of cacao". The tumpline is one of his symbols (Thompson 1960:76). The object held by the ascending figure may be the same form of torch as shown in the hands of God M in the Madrid Codex (see Thomspon 1960; Figures 4:13-4:22).

The scene is remarkably like a Mayan ritual, a fixed feast, known to take place in Muan, the fifteenth month of the year. The ceremony was held directly in the cacao plantations to envoke the intercession of the rain god and the god of commerce and included the sacrifice of a dog (Payon 1936:15-16). The propitiation of the gods associated with the cacao crop was an important event in the lowlands.

The two significant points of the sculpture for this study are that the scene depicts a fruiting cacao tree and a ritual which, in other lowland areas, is known to take place in the plantation. But what evidence is available from the field systems themselves?

Opposite Gutiérrez Zamora, on the south bank of the Rio Tecolutla, there are the vestiges of an unmapped field system (Figure 4:10). Today much of this area belongs to the small congregación of Cacahuatal. The Nahuatl term for cacao is cacahuatl. Cacahuatl can be translated as the "planting of cacao" or "place of cacao" and although the name cannot be traced in documentation further back than the nineteenth century, local oral history ascribes it to antiquity.

Apart from the possibility of an enduring place name there is far more dramatic evidence from System C on the north bank (see Figure 4:3). Here in the process of a botanical inventory of the Santa Luisa site an unusual specimen which proved to be Engallagema latifolia, or "wild cacao" was encountered. It is a rare tree in coastal swamps. A close examination of the 3 hectare forest produced a count of more than one hundred "wild cacao" trees in different stages of growth. This unusually large concentration may represent a residual population descendent from earlier large-scale Pre-Columbian plantings. Over one hundred informants from the Tuxpan, Tecolutla, and Nautla catchments were shown a large sample of leaf cuttings and fruit (Figures 4:11 and 4:12) and only two, from different areas, could identify the specimens; both stating they had only ever seen one example of the tree before.

"Wild cacao" is not cacao but rather an entirely different species which was sometimes planted in cacao plantations. The Nahuatl name for the tree is quauhcapatlachtli. This fruit, of similar form but less value, would be prepared in the same way and sometimes mixed with real cacao (Payón 1936:12). Its surprise appearance on the vestiges of the Santa Luisa irrigation

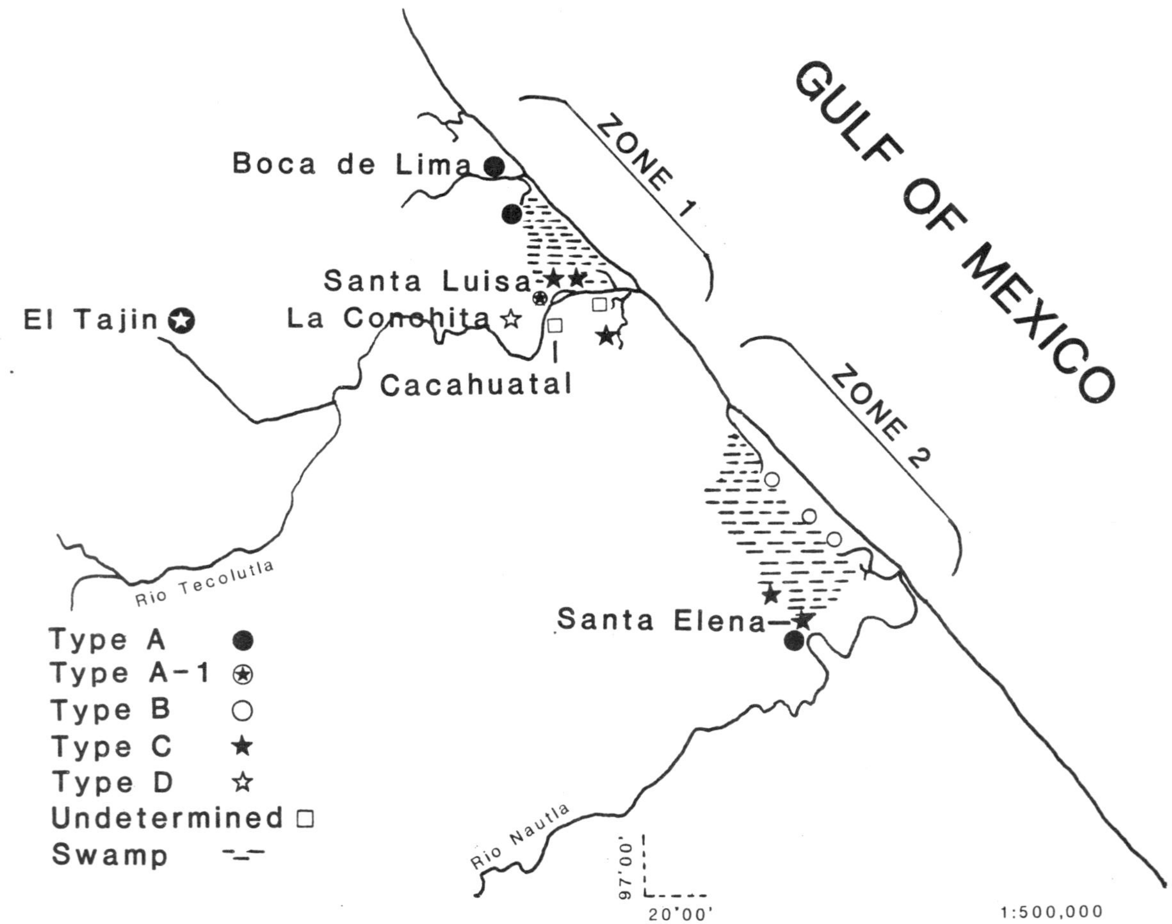


Figure 4:10 Field systems in the Tecolutla (Zone 1) and Nautla (Zone 2) drainages

Figure 4:11 Leaves and fruit pod of Engallagema latifolio from System C at Santa Luisa, Veracruz

Figure 4:12 Fruit pods of Engallagema latifolio from System C at Santa Luisa, Veracruz

system suggests the possibility that a portion of the system may have been originally planted with both wild cacao and cacao. Almost any cacao would have had a high commercial value and would certainly have been controlled by the Tajín state. Again, all these large-scale systems must be considered within the greater social context which controlled the distribution of the products derived from such technologies.

The reality of the utilization of the systems at Santa Luisa must lie in the cultivation of various types of crops (Figure 4:13). Nevertheless, sufficient maize to support the large population of Santa Luisa could probably be grown on roughly a third of the area covered by the north bank systems at the site. Most of the other two thirds were free to produce appropriate tribute or trade crops. This ratio may increase even more if we move beyond this site, albeit the most important in the lower Tecolutla drainage system, to consider the immediate surrounding area. Here too we find large field systems, but in most cases associated with much smaller sites than Santa Luisa.

Aerial observations have indicated numerous other irrigated areas on an axis stretching northward from Santa Elena past Santa Luisa to the minor drainage systems known as Boca de Lima (Figure 4:10). The area designated as zone 2 is geologically similar to that called zone 1 in that it represents a silted and enclosed embayment of essentially the same date. It is larger, however, and had a greater total area under cultivation when the systems were in use.

As discussed, both Santa Luisa and Santa Elena have block-like units of canals, averaging 3.2-3.9 hectares in size, forming a checkerboard pattern about major feeder canals. The Santa Luisa system and quite probably its counterpart at Santa Elena avoid the problem of fluctuating salt and brackish water by actually moving water from one or more fresh water sources through long canals descending gently across fertile river and/or estuary banks. The fresh water sources now have negligible flows due to deforestation, silting, and increased evaporation. Small house mounds, probably for canal maintenance, are found among the blocks.

These two major systems certainly grew in a series of steps. The largest, Santa Luisa, appears to have had at least three periods of growth. The first diverted fresh water tributaries of the Rio Tecolutla to form a rectilinear criss-cross pattern of canals; these were, on average, smaller than later additions. The area occupied approximately 0.6 square kilometres of land. Cores show that a former channel of the river became silted here following the Arroya Grande phase (ca. 300-1 B.C.) (Lentell 1975). The date of this system is uncertain but it may be as early as the Tecolutla (ca. A.D. 1-300) or Cacahuatal Phases (ca. A.D. 300-600). The later additions

<u>Figure</u> <u>4:13</u> Reconstruction of a portion of a downstream section of the systems at Santa Luisa. Caretaker's house is shown on left and one of the main feeder canals on right. The centre foreground has corn fields in various stages of growth with interspersed beans, chiles and squash. Cotton is in the distance as are the cacao groves. Paintings by Henry Sorenson under the direction of the author

follow the standard pattern and, based on preliminary ceramic and geologic evidence, date to the Isla A (ca. A.D. 600-900) and Isla B (ca. A.D. 900-1100) phases. These later sections occupy ca. 2.5 and ca. 2.65 square kilometres respectively. Both show considerable organization and engineering skills as the total natural elevation difference of the system is unlikely to have exceeded 1.5 metres from source to the outflow in the north and east. They may even have required the construction of a dam across the major stream flowing into the swamp which formed in the silting embayment. A dam has been observed but not explored archaeologically; it would have raised the water level by up to two metres in the northwest of the system. The same area also has what may be a kilometre long earth causeway connecting the northern portion of Santa Luisa with its Epi-Classic satellite site of La Isla.

The Santa Luisa system must be considered in the context of the other intensive agricultural areas surrounding it. These include the system of catchment terraces found among the sharp sandstone hills to the west (Wilkerson 1973, 1974). Here field fertility was increased by intercepting humus run-off behind catchment-like stone dams placed astride the small valleys between hills. There is some indirect evidence of deliberate deforestation to increase topsoil accumulation. The total extent of this terrace area is unknown but it is potentially up to fifteen square kilometres.

In the study area type A-1 (see Table 4:2) is found at Santa Luisa in the oldest of three identified segments. Chinampas are found at two locations near Boca de Lima and several near San Rafael. Raised fields are visible in several locations on the lee side of the present coastline adjoining a freshwater swamp in zone 2 (Figure 4:10). Irrigated fields are found at Santa Luisa in the two larger and more recent sections and at Santa Elena. Catchment terraces are presently only known in the area west of Santa Luisa but were probably quite common in the lower reaches of several nearby rivers.

The technology of the intensive agricultural systems in the Santa Luisa area appears to be a combination of lowland and high-land approaches. The chinampas (A and A-1 on Table 4:2), irrigated fields, catchment terraces, and hill-terraces, maybe of highland origin although a wide variety of systems are known to exist in the Maya lowlands (Harrison and Turner 1978). Raised fields are certainly a lowland development. Adaptation of the highland forms, however, does not imply exact duplication. The ingenious use of fertile deltaic river alluvium while avoiding salt and brackish water in irrigated fields is a good example of technological adaptation to a new environment.

The implications for the nature of the centralized El Tajín state are many. In fact, it is not improbable that El Tajin was more centralized and highly structured than

| Type | Designation | Definition |
| --- | --- | --- |
| A | Chinampa | Intensive cultivation in raised artifical earth blocks in a lake, lagoon, or water-course |
| A-1 | | A modified watercourse such as a river or stream expanded through numerous, basically parallel canals, to utilize alluvial terrace soils; cultivated areas frequently form small rectangular blocks |
| B | Raised Field | Prepared, raised earth embankments in parallel patterns with low points between for water collection and run-off, usually without major feeder and distributary canals |
| C | Irrigated Field | Grid-arranged fields, often raised and irrigated by diverted watercourses in feeder and distributary canals that carry, or potentially can carry, water throughout the year |
| D | Catchment Terrace | A stone fronted terrace in small valleys to catch and hold seasonal and induced soil run-off |
| E | Check Dam | A dam placed across hillside ravines to prevent soil run-off |
| F | Weir Dam | A dam placed across drainage channels to catch run-off |
| G | Hill Terrace | A terrace perpendicular to the slope of a hill or mountain to catch run-off and provide a level area for cultivation |
| H | Irrigated Terrace | A hill terrace with a ditch or canal providing water to the cultivated area |

Table 4:2 Provisional Classification of Intensive Agricultural Systems in Veracruz

its Mayan contemporaries. For the moment, it should be indicated that the extent and complexity of the sytems in the two study zones suggest a highly organized control institution. The nearly contiguous nature of the systems and their contemporaniety imply the presence of specialists or engineers who designed the systems around the economic requirements and hydrological limitations of each zone. The general impression is that the systems may have been planned and executed as macro-units. Site distribution about and within the systems also suggests special administrative points and the use of major canals for canoe transport.

## Other systems in the North-Central Gulf Area

There are several unmapped systems along the southern margin of the Río Nautla. Further south is the site of Las Higeras on the Río de Colipa, not far from Vega de Alatore. Just to the north-west of the site are Laguna Chica and Grande. Vestiges of field systems are visible along the southern side of these lagoons.

A careful study of this drainage system might show a similar relationship between Las Higeras and the nearby major site, Aparicio, as between Santa Luisa and El Tajín. Aparicio appears to be a major Late Classic and Epi-Classic site within the greater Tajín culture sphere. It may have been the capital of a major province paying tribute to El Tajín or, possibly have been an independent unit sharing Tajín culture.

Westward in the same drainage system as Las Higeras, on the edge of the Sierra de Chicoquiaco, is Colipa. Sixteenth century documentation (Paso y Troncoso 1905, V:115) suggests some form of irrigation here at the end of the Postclassic Period.

North of the Río Tecolutla there are additional systems. Ponciano Ortíz (personal communication 1977) encountered canals of an undefined configuration in the marshlands east of Postclassic Tabuco, near the mouth of the Río Tuxpan. Air reconnaissance along the coast in the 1970's also suggested the presence of two systems near the mouth of the Cazones River, one near the southern extremity of the marsh by the Tuxpan River, just south of Tamiahua and one immediately to the north of the same town. Undoubtedly there are many more locations to be encountered, particularly along fresh water creeks on the fringes of marshland.

The nature of these northern systems are likely to be similar to those of the Tecolutla and Nautla, although at first glance most appear to be smaller in extent. The majority, on the basis of the most preliminary aerial examination, correspond to raised field configurations of the type B variety, and a few to chinampas (type A-1).

The systems near Tamiahua border the upper limits of the North-Central Veracruz sub-culture area (Wilkerson 1973). Although it is possible that their dates are similar to those in the centre of the area the likelihood, given the nature of a south to north movement of cultural florescence and social centralization at the end of the Classic and during the Postclassic, is that they will be later (see Wilkerson, n.d.; 1981). Only field reconnaissance and excavation can confirm this supposition.

Ceramics, architecture, and art during the Classic Period reflect the considerable influence of El Tajín on this region. It is possible that at some point, probably during the Epi-Classic (ca. A.D. 900-1100), that it was brought under the direct control of this dynamic metropolitan centre. In fact, the important conquest celebrated by the ruler 13 Rabbit, and depicted on a carved column at El Tajin, may represent the subjugation of neighbouring portions of the Huasteca (Wilkerson 1980, in press).

Regardless of the role of El Tajín in the northward spread of intensive agricultural systems the Postclassic Period saw the rise of a series of militaristic kingdoms in the same region. Their success and relatively high population density suggest ample food availability that goes beyond simple swidden agriculture. There is a hint of this in the major surviving Pre-Columbian source for the region, the <u>Lienzos de Tuxpan</u>.

These maps, published by Melgarejo Vivanco (1970), date to both the period just before the Conquest and to the nineteenth century when two were copied by a local Italian artist. One of the earliest maps depicts the fifteenth century conquest by Tuxpan of the region between the Tuxpan (or Pantepec) and Tecolutla Rivers (Wilkerson 1980:203). Near certain place glyphs (Figure 4:14) are the standardized depictions of a plant/tree with a bulbular root or base.

Exactly what is portrayed is unclear but the bulb base and bifurcated stalk-top is highly reminiscent of freshwater reeds or the aquatic "<u>Lirio</u>". Both grow, even clog, slow-moving fresh water bodies today. Irrespective of the precise botanical identification, the locations suggest a rebus symbol or ideogram for some form of intensive agricultural system. Shown on the glyph are: 1) the marsh east of Tabuco, perhaps Tumilco; 2) the southern fringe of the marsh just landward of the coast; 3) a site near the island centre of Isla Farallón on the Cazones River not far from the river's mouth; 4) Santa Luisa, on the north margin of the Tecolutla River; 5) Ocelotepec; and 6) Tlaltizapan. The first four areas have known or probable field systems while Ocelotepec and Tlaltizapan have never been surveyed. The presence of these systems and their produce would certainly increase the desirability of conquering the region, and explain their

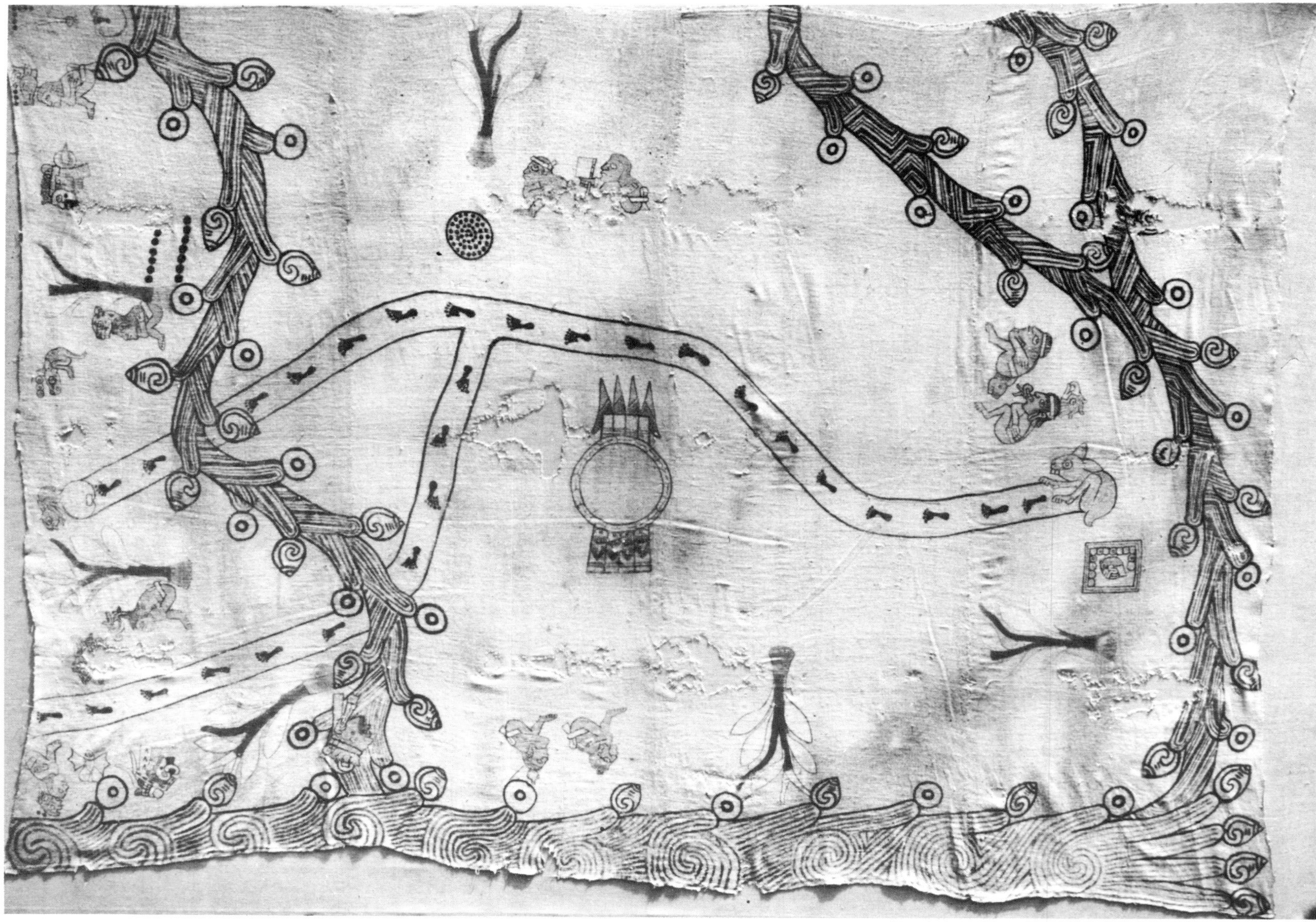

Figure 4:14 The "local map", one of the Lienzos de Tuxpan. Photo by David Hiser

inclusion on the map.

## Systems in the North Gulf Area

Shown in Figure 4:15 are some possible sites for field or terrace systems in the northern and western extremes of the Laguna de Tamiahua and the Río Panuco. These sites have been observed from the air but they have not been ground checked; in many cases they occur in areas of modern disturbance. Future field work will be required to confirm or eliminate the possible systems indicated. All the locations, however, are consistent with the hydraulic conditions appropriate for raised fields.

In terms of total distribution it is envisaged that the systems may not extend much further north than the region around the town of San Diego near the limits of the Sierra Otontepec. The North Gulf area apparently did not have highly centralized social units until relatively late in Pre-Columbian times and large scale systems may not have been built. Again future research will have to address this problem, but hydraulic agriculture can never be adequately understood without taking into account social centralization and the cultural parameters of an area.

## Systems in the South-Central Area

A very elaborate set of hill terraces are found on the south side of the mountain fortress of Quiahuitztlan (Wilkerson 1979) while others may occur at Plan de las Hayas, Cerro de la Morena and Xiochimalco. Similar terraces may also exist along the now barren canyon of the Antigua River upstream of Ocolapan not far from Rinconada. This area was once known as Izcalpan and was one of the very few sites in Veracruz claimed directly by Cortés. This selection, plus the fact it may have been utilized for sugar cane planting (Cortés 1528:396), suggests it must have looked very fertile at the time of the Conquest.

Zempoala, as noted earlier, almost certainly had irrigated fields. Other towns, Tepetlan and Atexca, in the Sierra west and north of Zempoala, may also have had irrigation systems (Kelly and Palerm 1952:99; Paso y Troncoso 1905 V:117,120). Still other possibilities are suggested in the same general region by glyphs and corn plants depicted in the early colonial Codex Misantla. Ixhucan, high in the Sierra Madre Oriental, is another strong possibility for a Postclassic system.

Close examination of the northern portion of South-Central Veracruz and the Sierra de Chicoconquiaco is likely to result in the location of additional systems, particularly varieties that were designed to transport water onto terraces and natural levees. However the rain shadow area of the coastal plain and many of the valleys of the sierra are in a drier zone and are not likely to have raised fields or systems to remove excess run-off.

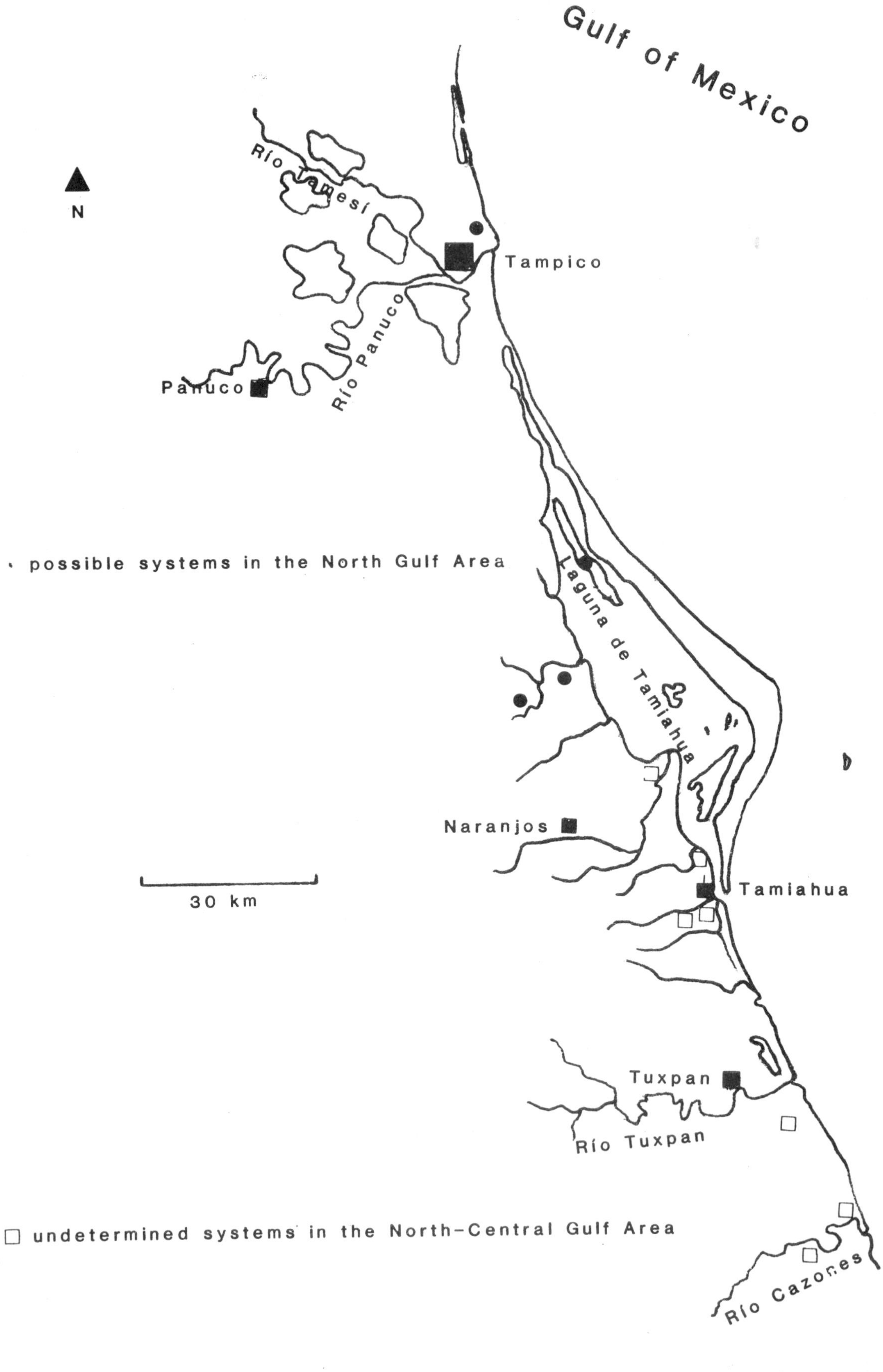


Figure 4:15 Possible field systems located from the air in the North Gulf culture area

Further south and nearer the coast however, in the San Juan drainage system, Alfred Siemens is investigating fields of this nature (see Chapter 3). From the air the western and southern margins of the Laguna de Mandinga shows features consistent with a raised field system. Undoubtedly other examples will be encountered, particularly along the margins of the Bay of Alvarado and the Papaloapan River with its major coastal tributaries the San Juan and Tesechoacan.

## Systems in the South Gulf Area

In spite of the fact that this area has many of the natural features appropriate for systems such as raised fields, none have ever been noted. Many of the fertile terrace soils, such as the seasonally flooded margins of the Coatzacoalcos or Tonala Rivers, could certainly have been utilized intensively. The lack of any intensive agricultural schemes suggests that there were not sufficiently large and centralized social units in existence at the time the technology was current. In other words, the presence of readily exploitable natural conditions, in the absence of well integrated social communities to command the necessary labour for construction and maintenance, would limit the scope of any hydraulic endeavours.

This is an intriguing possibility for a region well within the core area of Mesoamerica. Unlike the North Gulf area, a labile frontier throughout its Mesoamerican history (Wilkerson 1981; n.d.), this area must be influenced by other factors.

During the Formative Periods the area formed a major portion of the Olmec "heartland" (Coe 1965); the dynamic centre of Mesoamerican culture at the time. By the Classic Period the focus of events, at least as far as the lowlands are concerned, had moved northward into the South-Central area and east, towards the Peten. The only excavated site in the area with a long chronology, San Lorenzo (Coe 1969, 1980), suggests abandonment at this time. In contrast Cerro de Las Mesas, to the north, and Palenque and Comalcalco to the east, reached their apogees at precisely this time.

The area is not devoid of Classic Period occupation, it appears, on the basis of very limited information, to have many scattered small sites on the ridges beyond the flood plains. As a corridor for migration and conquest, as well as a buffer between the more dynamic lowland centres to the north and west, the stability of social units in the area must have been precarious at best. Large-scale intensive agricultural systems would have been difficult to plan, execute, and then maintain in such a chaotic milieu.

The Postclassic may, in spite of widespread migration throughout much of the Mesoamerican lowlands, have brought

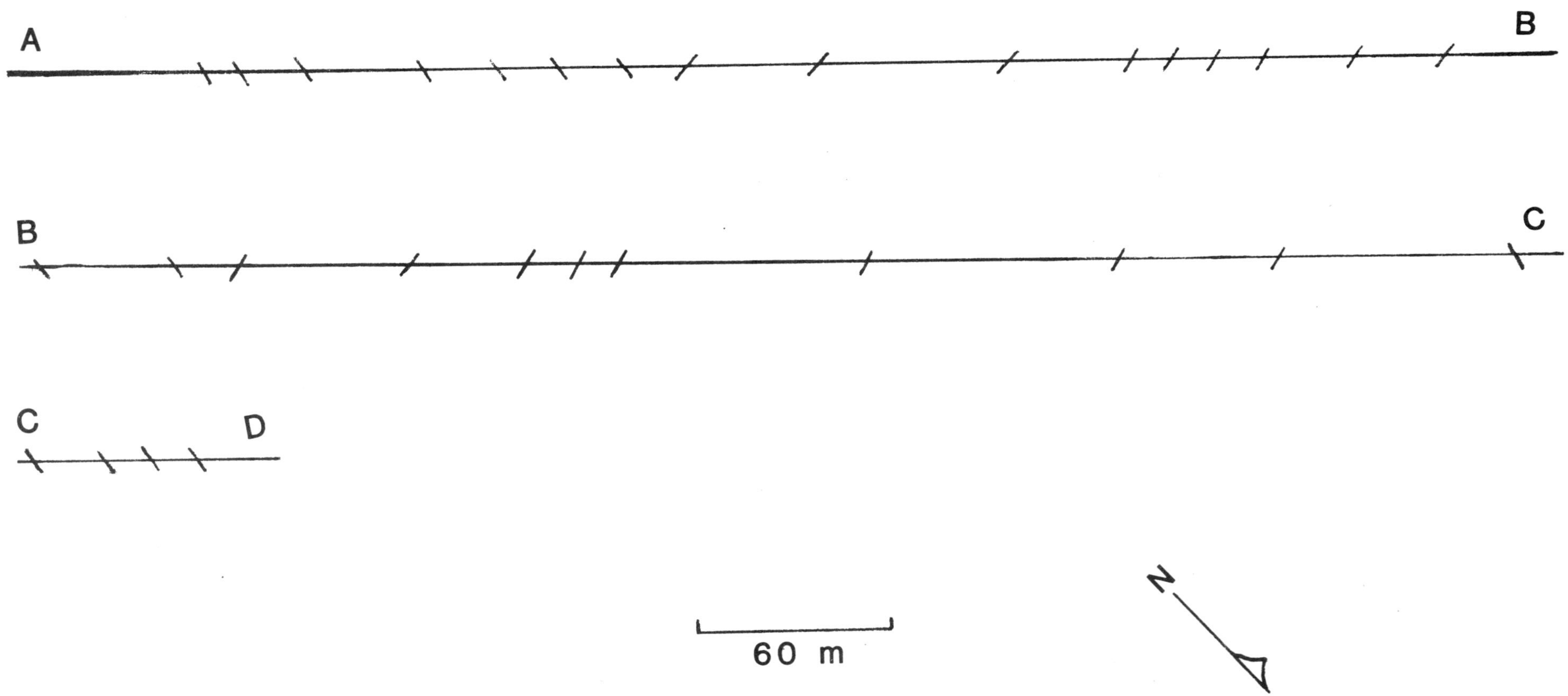


Figure 4:16 Canal directions and groupings along a 1020 metre transect of System C at Santa Luisa

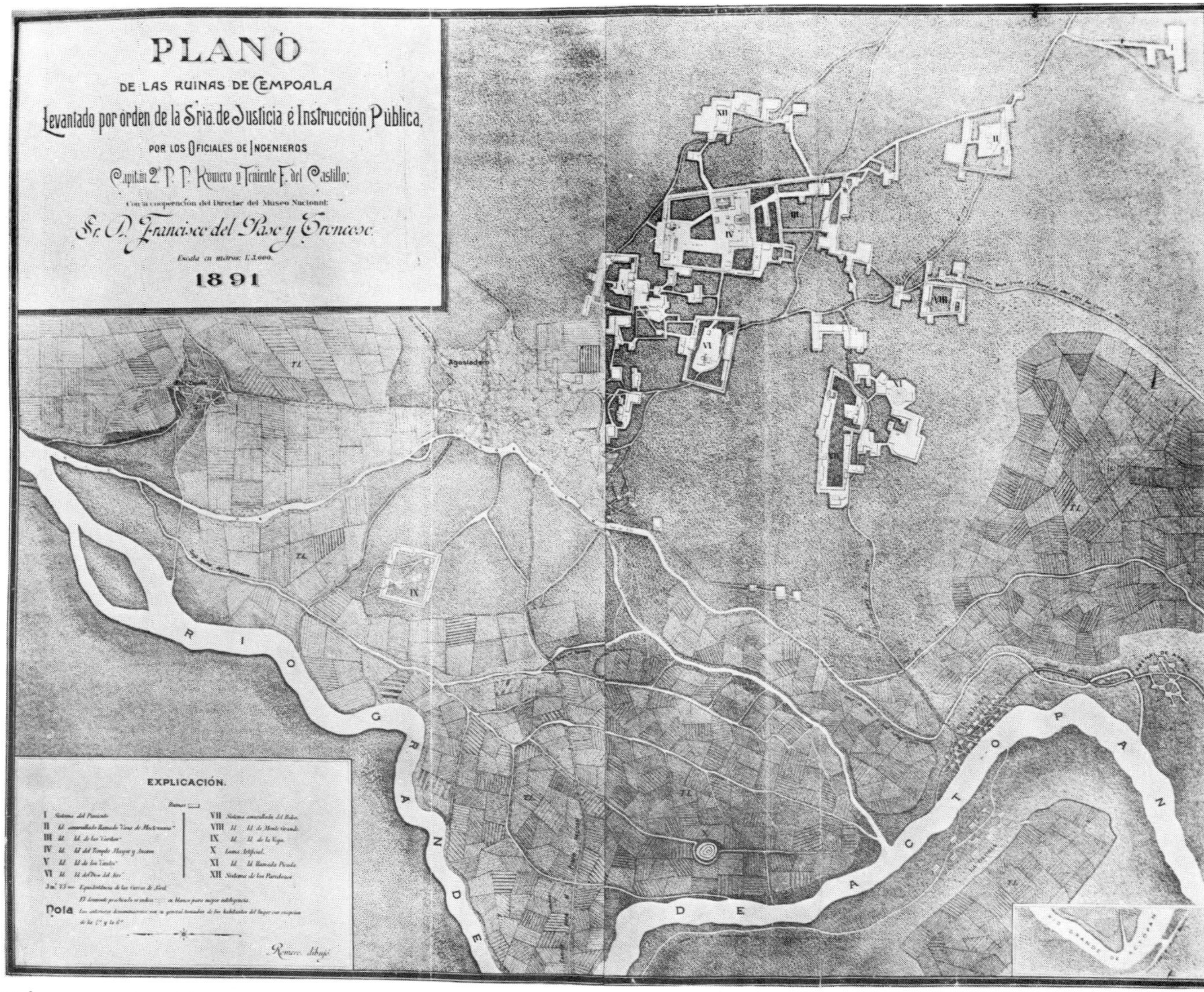


Figure 4:17 The ruins of Zempoala in 1891 with the remanents of one of the irrigation systems in the foreground. Map from Paso y Troucaso

a slightly greater stability to certain portions of the area. Coe (1980) recorded the greatest activity since the Formative at San Lorenzo between A.D. 900 and 1100 Later the region became a province of the Aztec tribute state. Very soon after the Conquest, Cortés, considering the region to be important economically, sent Gonzalo Sandoval to found the Villa de Espiritu Santo.

While we should bear in mind that the absence of known field systems and terraces in the area must in part be an artefact of limited investigation, social factors of the Classic Period suggest large-scale systems were not common. Postclassic systems might be more abundant but are unlikely to be as numerous as those known for the areas to the north.

## Conclusions

From this survey of evidence of intensive agriculture in ancient Veracruz we can draw some conclusions and hypotheses to be tested by future investigations:

1. Intensive agricultural systems have existed on alluvial terraces in Veracruz since the Early Classic Period and perhaps from the Proto-Classic as shown by preliminary evidence at the site of Santa Luisa;

2. A number of irrigated field systems made use of fresh water tributaries, diverting them onto the terraces of tidal rivers;

3. The systems in the North-Central Gulf area appear to have been extended rapidly during the Late Classic and Epi-Classic Periods, giving the appearance of well-planned large-scale units fitting into an overall regional plan and engineering style;

4. Initial examination of production potential suggests that the systems associated with the Tecolutla and Nautla rivers could produce more food and fibre than the estimated needs of the archaeologically verifiable population. It is hypothesized that excess produce grown in up to two thirds of the agricultural area, were used for the tribute or taxation due to the major metropolitan centre, El Tajín, in the middle reaches of the Tecolutla;

5. While portions of these systems were devoted to corn, beans, squash, and chile production for local consumption the "tribute" portions may have consisted of crops such as cotton and cacao. Direct evidence for cacao comes from El Tajín and Santa Luisa where large concentrations of "wild cacao" grow on the uncleared portions of ancient fields;

6. Classic Period raised field systems are common along the North-Central Gulf coast as far north as the southern margins of the Laguna de Tamiahua. Chinampas

are also frequent, and with the exception of one subtype, may be typical of the Postclassic Period culture. Some, in fact, may be documented in contact period accounts such as the Lienzos de Tuxpan. Catchment terraces are also to be found in this area;

7. In the North Gulf area a number of possible locations for systems occur near the Laguna de Tamiahua and the Río Panuco. However, it is hypothesized that direct examination may only verify a few because less centralized social organization characterised the area in the frontier-like Huasteca until Postclassic times;

8. The South-Central Gulf area has numerous Postclassic irrigated and terrace forms mostly of Postclassic date. A few Classic Period raised fields have been identified and many more are likely to be encountered once survey programmes in the area increase;

9. No systems have yet been found in the South Gulf area. It is hypothesized that the insecure social environment of the Isthmus as a major migration route, and as a buffer between dynamic but differing lowland groups, prevented the rise in the Classic and Early Postclassic periods of sufficiently powerful social units to implement large systems until near the time of European Contact;

10. The intensive agricultural systems in ancient Veracruz show diversity and complexity which both parallel and surpass the reported systems for the Classic Maya region. The large-scale irrigated and raised field systems in the region of El Tajin foreshadow the contributary context of intensive low-land agriculture in the Postclassic that would help sustain the Aztec tribute state. Bernal Díaz's amazed observation was correct, Veracruz was "green with vegetation", but the question must always be asked ... whose garden was it?


## Acknowledgements

The research in Veracruz was undertaken in collaboration with, and under the supervision of the Instituto Nacional de Antropología e Historia which is thanked for the corresponding permits. The splendid assistance and encouragement of the late Arq. José García Payón of INAH is gratefully acknowledged. Since 1973 the investigation has been undertaken with the aid of grants from the National Geographic Society whose generous support is greatly appreciated.


## Bibliography


CODEX, M. 1979, Coleccion de Mendoza or Codice Mendocino, Fascimile Fototipico de Don Francisco del Paso y Troncoso, 1925, edicion de Editorial Cosmos, Mexico.

COE, M.D. 1965, The Olmec style and its distribution, In, Handbook of Middle American Indians, vol.3, edited by R. Wauchope, University of Texas Press, Austin, 739-775.

COE, M.D. 1969, The archaeological sequence at San Lorenzo, Veracruz, Mexico, presentation, Society for American Arcaheology, Milwaukee.

COE, M.D. and R.A. DIEHL 1980, In the Land of the Olmecs, 2 Volumes, University of Texas Press, Austin, Volume 1, 416pp, Volume 2, 198pp.

CORTÉS, H. 1528, Memorial de servicios y conquistas hecho para el emperador a solicitud de este, In, Cartas y Documentos, Editorial Porrua, 1963, 395-399.

CORTÉS, H. 1979, Cartas de Relación, Editorial Porrua, Mexico City, 331pp.

DÍAZ DEL C.B. 1974, Historia Verdadera de la Conquista de la Nueva España, Editorial Porrua, Mexico City, 700pp.

GALINDO y VILLA, J. 1912, Las Ruinas de Cempoala y del Templo del Tajin, notes of Francisco del Paso y Troncoso, Mexico, 121pp.

GARCIA PAYON, J. 1936, Amaxocoatl o Libro del Chocolate, Toluca, Escuela de Artes, Toluca, Mexico, 122pp.

GARCÍA PAYÓN, J. 1966, Prehistoria de Mezoamerica, Excavaciones en Trapiche y Chalahuite, Veracruz, Mexico, 1942, 1951, y 1959, Guadernos de la Facultad de Filosofia, Letras y Ciencias, Num.31, Xalapa, 204pp.

GIBSON, C. 1964, The Aztecs Under Spanish Rule, Stanford University Press, Stanford, 657pp.

HARRISON, P.D. and B.L. TURNER II, (editors) (1978), Prehispanic Maya Agriculture, University of New Mexico Press, Albuquerque, 414pp.

KELLY, I. and A. PALERM 1952, The Tajín Totonac, Part I, History, subsistence, shelter and technology. Pub.13, Institute of Social Anthropology, Smithsonian Institution, Washington, 369pp.

LENTELL, R.L. 1975, Depositional History of the Rio Tecolutla Estuary, Mexico, Ph.D. thesis, University of Florida, Gainesville, 44pp

MELGAREJO VIVENCO, J.L. 1970, Los Lienzos de Tuxpan, Editorial la Estampa Mexicana, Mexico City.

MOLINA, ALONSO DE 1977, Vocabulario en Lengua Castellana y Mexicana, Editorial Porrua, Mexico, 163pp.

PALERM, A. 1955, The agricultural bases of urban civilization in Mesoamerica, In, Irrigation Civilization, edited by J.H. Steward, Pan American Union Social Science Monographs 1, Washington, D.C., 28-42.

PASO Y TRONCOSO 1905-06, Papeles de Nueva Espana, seguna sene, geografia y estadistica, Mexico, 7 volumes, 1913pp.

SCHMIDT, P.J. 1977, Un sistema de Cultivo Intensivo en la Cuenca del Rio Nautla, Boletín del Instituto Nacional de Antropología e Historia, Epoca III, Mexico City, 20, 50-60.

STREBEL, H. 1884, Die Ruinen von Cempoallan in Staate Veracruz und Mitteilungen uber die Totonaken der Jetstzert, In, Abhandlungen aus dem Gebiele der Naturwissenchaften, VIII, Hamburg.

THOMPSON, J.S. 1960, Maya Hieroglyphic Writing, University of Oklahoma Press, Norman, 500pp.

TORQUEMADA, F.J. DE 1943, Monarguia Indiana, tomo 1, Editorial Salvador Chavez Hayhoe, edicion facsimile, Mexico, 885pp.

WILKERSON, S.J.K. 1972, Ethnogensis of the Huastecs and Totonacs; Early cultures of North-Central Veracruz at Santa Luisa, Mexico, Ph.D. dissertation, Tulane University, 1004pp.

WILKERSON, S.J.K. 1973, Report on the 1973 season of the Florida State Museum-National Geographic Society Cultural Ecology Project, Gainesville, 14pp.

WILKERSON, S.J.K. 1974, Cultural subareas of Eastern Mesoamerica, Primera Mesa Redonda de Palenque, Robert Louis Stevenson School, Pebble Beach, California, 89-102.

WILKERSON, S.J.K. 1974, Report on the 1974 season of the Florida State Museum-National Geographic Society Cultural Ecology Project, Gainesville, 22pp.

WILKERSON, S.J.K. 1975, Report on the 1975 season of the Florida State Museum-National Geographic Society Cultural Ecology Project, Gainesville, 25pp.

WILKERSON, S.J.K. 1976, Haustec presence and cultural chronology in North-Central Veracruz, Actes du XLII Congres International des Americanistes, IB-B, Paris, 30-47.

WILKERSON, S.J.K. 1976, Report on the 1976 season of the Florida State Museum-National Geographic Society Cultural Ecology Project, Gainesville, 92pp.

WILKERSON, S.J.K. 1978, Report on the 1972-1978 seasons of the Florida State Museum-National Geographic Society Cultural Ecology Project, Gainesville, 71pp.

WILKERSON, S.J.K. 1979, Mesoamerican influences on the Northern Gulf Coast, Lecture, Tulane University, New Orleans, December.

WILKERSON, S.J.K. 1979, The 1979 season of the Florida State Museum-National Geographic Society Cultural Ecology Project, Gainesville.

WILKERSON, S.J.K. 1980, Man's eighty centuries in Veracruz, National Geographic Magazine, 158, (2), 202-231.

WILKERSON, S.J.K. 1981, The Northern Olmec and Pre-Olmec frontier on the Gulf Coast, In, The Olmec and Their Neighbours, edited by M.D. Coe and D. Grove, Dumbarton Oaks, Washington, 181-194.

WILKERSON, S.J.K. 1983, In search of the mountain of foam: human sacrifice in Eastern Mesoamerica, In, Human Sacrifice in Pre-Columbian America, edited by J. Soustelle, Dumbarton Oaks, Trustees for Harvard University, Washington, D.C, 43-73.

Chapter 5

# *ANCIENT RAISED FIELD AGRICULTURE IN THE MAYA LOWLANDS OF SOUTHEASTERN MEXICO

Stephen R. Gliessman, B.L. Turner II, F.J. Rosado May and M.F. Amador

## Abstract

The lowland floodplains of tropical southeastern Mexico have posed a great obstacle to agricultural development in recent times. Studies of Pre-Hispanic land use of such areas can provide invaluable information concerning potential design and management of agriculture for the future. One particular system studied in detail is reported here. In southern Quintana Roo, Mexico, a depression (bajo) of several thousand hectares has recently been cleared for commercial rice production, revealing an intricate network of canals and raised platforms. Soil description and analysis, as well as topographic surveys, indicate past human intervention. Past management possibilities, present-day traditional use of such wetlands and future applications will be discussed.

## Resumen

Los llanos de inundación del sureste tropical de México han presentado un gran obstáculo al desarrollo agrícola en tiempos recientes. Estudios del uso de la tierra pre-hispánico de tales areas pueden proporcionar valiosa información sobre el diseño y manejo potencial de la agricultura en el futuro. Aquí se informa sobre un sistema particular que se estudió en detalle. En el sur de Quintana Roo, México, un bajo de varios miles de hectáreas se ha despejado recientemente para la producción comercial de arroz, revelando una red detallada de canales y plataformas elevadas. Tanto la descripción y análisis de los suelos como los levantamientos topográficos indican intervención humana en el pasado. Se discutirán posibilidades de manejo en el pasado, el uso tradicional actual de estas tierras húmedas y sus aplicaciones futuras.

## Introduction

Recent studies have demonstrated that the prehistoric Maya of the Yucatan peninsula developed several types of technologies that enabled them to cultivate wetlands (Puleston 1977; Turner and Harrison 1981; Siemens 1982). Initial observations were made along water courses (Siemens and Puleston 1972), but were subsequently extended to other low-lying terrain throughout southern Quintana Roo and northern Belize (Turner 1974; Harrison 1977; 1978; Siemens 1978; n.d.; Turner and Harrison 1981; 1983), and perhaps to Peten, Guatemala, (Adams 1980; Adams

---

*Paper presented at Conference, "Prehistoric Intensive Agriculture in the Tropics", organized by the Australian National University, Canberra, Australia, August 1981.

et al. 1981) (Figure 5:1). Observations of ground patterns suggested that large areas of wetlands had been devoted to some kind of cultivation, presumably input and output intensive agriculture, that relied on systems of canals/drains, raised platforms or other structures that permitted water manipulation.

Despite the number of observations of ground patterns thought to be relics of wetland agricultural systems, ground studies have been limited to small networks of fields along the Rio Candelaria, Campeche (Siemens and Puleston 1972), and in northern Belize at Lamanai (Lambert and Arnason Chapter 6), Cerros (Scarborough 1980), Albion Island (Puleston 1977; Siemens 1982) and Pulltrouser Swamp (Turner and Harrison 1981; 1983). Most of the evidence of field-canal morphology, associated cultivars, dates, and so forth come from the latter two studies on systems that include not much more than 300 ha of wetlands. Indeed, Siemens (1982) estimates that northern Belize contains about 3,200 ha of fields, although remote sensing work suggests that wetland manipulation may have encompassed a larger area (Adams et al. 1981).

In contrast to northern Belize, southern Quintana Roo contains a vast network of depressions (bajos), many of which have ground patterns indicative of fields and canals (Turner 1974; Harrison 1977; 1978; Siemens 1978; 1982; Turner and Harrison 1981). Indeed, a conservative estimate of the area of bajos with ground patterns in Quintana Roo is 40,000 ha (Turner and Harrison 1983). Such large zones of potential wetland fields are critical to interpretations of Maya prehistory and has sparked controversy (Puleston 1978; Sanders 1979; Turner and Harrison 1979). This report presents the result of ground surveys and analysis of one projected wetland system in a large bajo complex of southern Quintana Roo.

## The Study Area

This study was carried out in the Bajo Morocoy, situated immediately northeast of the town of Nicolas Bravo in the southeastern part of the state of Quintana Roo, Mexico. The region is part of the flat, karstic, low-lying coastal margin of the limestone shelf forming the Yucatan Peninsula (Figure 5:1). Annual rainfall, registered at Nicolas Bravo (11 year record), is 1,323 mm, with 76 percent occurring between May and October, during the wet season (Lopez et al. 1977). Cyclones may occur between August and October and these occasionally cause serious flooding, as can rare storms from the north (nortes) during November and December. The driest month is April. The mean annual temperature is 25.7°C. The natural vegetation of the depressions is semi-deciduous to deciduous forest, 4-8 metres tall, with dominant species such as *Haematoxylum campechianum*, *Metopium brownei*, *Cameraria latifolia*, *Plumeria rubra* f. *acutifolia*, and *Achras zapota* (Pennington and Sarukhan 1969; C.P. Cowan, personal communication). The particular depression

Figure 5:1 Map of the area of study, Bajo Morocoy, near Nicolas Bravo, Quintana Roo, Mexico

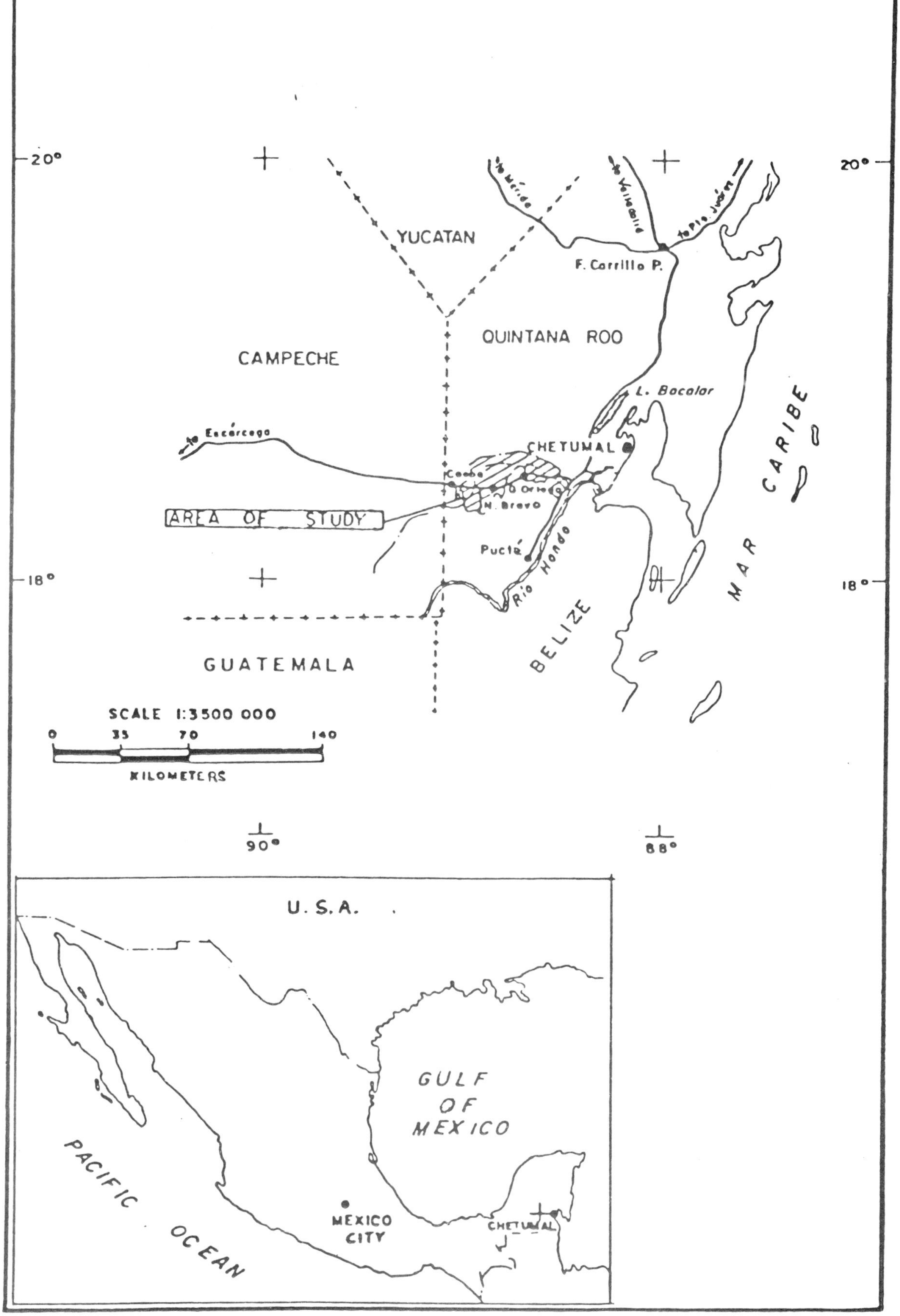

studied is at an elevation of 80-85m above sea level, forming in total an area of approximately 15-20,000 hectares. There is very little slope in the depression, topographic studies done by the Secretaria de Agricultura y Recursos Hidraulicos (SARH) showing less than a 0.3 percent gradient.

The depression (or *bajo*) has slow interior drainage, resulting in various degrees of seasonal inundation. The intermittent Rio Ucum arcs around the *bajo*, although the effect of that water course in the *bajo* is uncertain. Soils in the *bajos* tend to be Mollisols (Lopez *et al.* 1977; Turner 1978), mostly of the haplaquoll variety, that are characterised by high levels of montmorillonite clay and surface conditions which are sticky and impermeable in the wet season. The soils are relatively deep (1-2m) and light to dark grey in colour. They are formed from the *in situ* weathering of the calcareous bedrock and the deposition of colluvium from slopes surrounding the *bajos*.

Because of the periodic flooding, the agricultural colonisation plan for the southeast of Mexico, administered by the Secretaria de Reforma Agraria (SRA), proposed that such *bajos* be used primarily for growing rice. An area of 2,400 ha was mechanically cleared in early 1979 in Bajo Morocoy and planted with the aid of SARH. Field preparation was with tractors, and all seeding, fertilizing and spraying was done aerially. Harvesting was done with combines. The intensity of rainfall during the first planting season (1979) was much greater than normal (exceeding 2,000 mm) with a peak occurring at the time the rice was ready to harvest (November-December). A large portion of the crop was lost, and less than half of the initial investment was recovered (SARH, 1980).

The Colegio Superior de Agricultura Tropical (Cardenas, Tabasco) approached the planning agencies concerning the study of present *bajo* conditions and cultivation systems, and SARH invited us to conduct studies at Bajo Morocoy. This *bajo* is part of a complex stretching from Nicolas Bravo to Ucum which is reported to contain approximately 34,600 ha of relic fields and canals. Indeed, such systems were photographed or observed on government aerial photographs of the *bajo* by Turner (1974), Harrison (1977) and Siemens (1978) before it was cleared for the present agricultural scheme. Aerial photographs of the cleared *bajo* taken by SARH showed the same place and pattern originally photographed by Turner (Figures 17.1 and 17.2 in Turner and Harrison 1978). We were able, then, to pinpoint on the ground the location approximately 33.6 km north of the Escarcega-Chetumal highway, a location in which SARH had excavated several trenches across a complex of field-like platforms and canal-like depressions. We were fortunate to arrive in time to examine the exposed material during the dry season of May 1980.

## Methods

In order to describe the differences observed in the microrelief of the *bajo*, 100m long transects were surveyed with a theodolite, with elevations taken every 5m along a total of 40 lines. Half of the transects were done across what appeared to be a major canal with raised surfaces on both sides, and the other half in an area of raised surfaces and a complex lattice-work of interconnected smaller canals. After determining elevation differences, the average horizontal and vertical dimensions of the platforms and canals were calculated.

SARH's excavations, using a backhoe, had opened trenches with dimensions of 1m wide, 1.5m deep and approximately 50m long. In both, the trenches passed through at least two platforms and one canal. One extended across a major canal and the other through an area of smaller canals. We described the soil profiles and collected soil samples at 4m intervals. Soil chemical analysis using standard methodologies employed by the ecology and soils laboratories at the Colegio Superior de Agricultura Tropical were undertaken. Samples were also taken from each soil horizon and placed in tares for gravimetric determinations of percent soil moisture after drying for 24 hours at 105°C. For all samples, equal amounts of soil were collected at the top, middle and bottom of each horizon, well mixed, and then duplicate samples taken for analysis. Total sample weights in the field exceeded 200g.

An additional visit was made to the site in Early September 1980 to make observations of soil moisture conditions and crop response for the second season's planting of rice at the peak of the wet season. Measurements were made on rice plants growing in the canals and on the platforms with the idea of using the plants themselves as indicators of soil conditions. At ten randomly chosen locations (on both canals and platforms), 50 x 50 cm plots of rice were cut to ground level. The total number of stems, average height (cm) per plant, and biomass per plot (dry weight after 48 hours at 70°C) were determined.

## Results

The differences in microrelief indicate a fairly regular pattern of elevated surfaces and interconnected depressions between them. The average height of the platforms is 35-45 cm higher than the depressions for the site across a large canal (site 1), with the same difference occasionally found in the areas of small canals, but usually the difference is not as great (site 2). The depth of the large canals is 25-30 cm lower at its lowest point than the area of smaller interconnected canals. The slope surveyed in the large canal is approximately 0.3 percent, indicating the relative flatness of the *bajo*.

Table 5:1 Height (cm) of platforms as compared to adjacent canals along two transects at two sites in the Bajo of Morocoy, near Nicolas Bravo. Readings taken at 5m intervals

Site 1

| N ⟶ S | 20.1 | 48.4 | 41.0 | 20.2 | 8.4 | 11.7 | 3.1 | 0 | 26.0 | 39.9 | 35.7 | 42.7 | 34.2 | 44.2 | 40.2 | 28.5 |
|---|---|---|---|---|---|---|---|---|---|---|---|---|---|---|---|---|
| | | platform | platform | | canal | canal | canal | canal | | platform | platform | platform | platform | platform | platform | |
| | | | | | | | water found at 1.5m | | | | | | | | | |

Site 2

| E ⟶ W | 30.8 | 43.9 | 44.1 | 18.1 | 20.1 | 10.8 | 10.3 | 30.9 | 21.0 | 0.3 | 0 | 7.0 | 2.0 | 19.0 | 15.8 | 6.3 |
|---|---|---|---|---|---|---|---|---|---|---|---|---|---|---|---|---|
| | platform | platform | platform | platform | | canal | canal | platform | platform | canal | canal | canal | canal | platform | platform | |

Two representative transects are presented in Table 5:1. Horizontal dimensions of the system are quite variable. The platforms are square to rectangular ranging from 10 to 25m wide. The area of the raised surfaces varies from 400 to 600 square metres. The several large canals which stretch across the cleared part of the *bajo* are up to 20m wide, whereas the interconnected secondary canals vary from 8-12m wide. All of the secondary canals connect with the principal canals.

Considerable soil differences were found between the platforms and canals (Table 5:2). Covering the entire surface of the *bajo* to a depth of 25.0 cm is a dark brown crumbly clay layer. Below this is another 15-20 cm layer, speckled grey-brown in colour, much more compacted than the upper layer. It is interesting that this layer extends 15 cm deeper in the large canal at Site 1. The third horizon occurs only on the platforms, being a very dark grey clay 20-25 cm thick, that is very distinct from the two layers above as well as the 40-50 cm thick very light grey clay that follows. The dark grey horizon is missing from the canals. The unconsolidated limestone (locally called *sascab*) underlying the entire area is light grey to white in colour, and contains considerable carbonate and gypsum. Rock is generally lacking from the upper horizons. The total depth of the soil layer is greater on the platforms than in the canals, especially in the area of the secondary canals (over 30 cm difference). These soils correspond to a humic gley category of the mollisol order (Jenny 1980).

Laboratory analysis of the soils points out some interesting contrasts (Table 5:3). As would be expected for this soil type, pH rises with depth, although the soil on the platforms is more acidic, especially the dark grey layer. Organic matter declines considerably with depth, but less rapidly on the platform, where the dark layer still retains more than 2.5 percent organic matter. Exchangeable phosphate is low in all horizons, but there is a twofold difference between the platforms and the canals. Total nitrogen is highest at the surface and drops off rapidly with increasing depth. The second, speckled horizon on the platform shows lower total nitrogen than the canal soil for the same horizon. Nitrate is high only near the surface, being greatest in the second horizon of the canals. Potassium shows the same trend as nitrate. Calcium and magnesium, on the other hand, increase with depth, reaching highest levels at bedrock, and show similar trends in both the canals and platforms.

Except for the surface horizon in the dry season, all others were very moist, both in the wet and dry seasons (Table 5:4). It is only for the first (0-25 cm) and second (25-45 cm) horizons that differences occur from one season to the next. Below 45 cm, moisture levels in the soil are quite high, 70 percent or more moisture being taken to indicate saturation. The low moisture level of

Table 5:2 Average depth of soil horizons in canals and platforms at two sites in the Bajo of Morocoy, near Nicolas Bravo, Quintana Roo, Mexico

| Horizon description | Site 1 Platform (cm) | Site 1 Canal | Site 2 Platform (cm) | Site 2 Canal |
|---|---|---|---|---|
| Dark brown, dry aggregates | 25.0 ± 5.0 | 25.0 ± 0 | 25.2 ± 0.35 | 25.0 ± 0 |
| Speckled grey-brown gley | 40.0 ± 7.6 | 60.0 ± 10.0 | 46.2 ± 2.3 | 45.0 ± 0 |
| Dark grey gley | 70.0 ± 12.6 | missing | 72.5 ± 26.2 | missing |
| Very light grey gley | 110.0 ± 0 | 100.0 ± 7.1 | 128.3 ± 22.5 | 96.7 ± 5.8 |
| Decomposed limestone | 110.0 | 100.0 | 128.3 | 96.7 |

Table 5:3 Soil analysis of samples from platforms and canals at the Bajo of Morocoy

| Horizon (cm) | pH | O.M % | $PO_4$ ppm | Total N % | $NO_3$ | Ca | Mg | K |
|---|---|---|---|---|---|---|---|---|
| | | | | | me/100g | | | |
| Surface 0-25 | 5.8 | 3.40 | 0.70 | .153 | 47.00 | 51.20 | 19.2 | .406 |
| Platform 25-45 | 5.8 | 3.55 | 2.79 | .091 | 31.25 | 53.90 | 18.4 | .389 |
| Platform 45-75 | 5.8 | 2.54 | 1.87 | .041 | 2.54 | 54.56 | 19.0 | .280 |
| Platform 75-100 | 7.4 | 1.58 | 2.17 | .021 | 0.29 | 64.20 | 20.5 | .193 |
| Canal 25-45 | 6.5 | 3.25 | 1.08 | .140 | 54.70 | 58.48 | 18.0 | .558 |
| Canal 45-105 | 7.1 | 1.26 | 0.96 | .034 | 4.60 | 58.98 | 19.6 | .278 |
| Sascab >105 | 7.3 | T | 0.78 | .025 | 0.50 | 78.26 | 23.2 | .225 |

Table 5:4 Gravimetric moisture determinations in soils collected during both the wet and dry seasons at the Bajo of Morocoy

| Horizon (cm) | | 15 May 1980 % moisture | 6 Sept. 1980 % moisture |
|---|---|---|---|
| Surface | 0-25 | 25.0 ± 4.2 | 52.4 ± 5.3 |
| Platform | 25-45 | 47.3 ± 2.8 | 64.1 ± 5.4 |
| Platform | 45-75 | 62.5 ± 5.9 | 65.7 ± 2.8 |
| Platform | 75-110 | 67.1 ± 5.2 | n.d |
| Canal | 25-45 | 66.4 ± 6.9 | 64.8 ± 3.2 |
| Canal | 45-100 | 70.8 ± 2.6 | 66.0 ± 1.8 |
| Sascab | >100 | 41.7 ± 3.1 | n.d |

the sascab material is due to the high rock content in the sample. Interestingly, despite reports of a deep water table (Lopez *et al.* 1977), water was encountered at a depth of 1.5m *in the* trench opened across the large canal at the end of the dry season. Below 45 cm no soil differences were found between the platforms or the canals, regardless of the season. When the study area was revisited again in the wet season, the development of the 1980 rice crop closely reflected the topographic differences between the fields and the canals. Best growth (to the point of flowering) is seen in all aspects measured for the plants growing in the canals (Table 5:5). The plants on the raised surface were dry and stunted, and probably would never flower. For the entire crop, harvest was to begin one and a half months after the data were collected.

## Discussion

The topographic patterning in the Bajo Morocoy, as shown by ground survey and soil analysis, supports the belief that such depressions were used by the ancient Maya for intensive agricultural production. The interpretations of our observations are diagrammatically presented in Figure 5:2. Apparently the dark grey soil layer (layer 3) found only on the platforms is the remnant of the ancient planting bed that was excavated from the lower canals/ditches and mounded up to form a planting surface which would be available for cropping even at the peak of the wet season. This soil layer probably formed the original surface of the *bajo*, which was excavated down to the parent material and mounded up, creating platforms that once stood 60-70 cm high during the time of use. The penetration of the "sascab" zone may have been a result of an attempt to obtain a suitable height between the top of the field and the maximal water level in the canal. The surface areas of the platforms, 400-600 $m^2$, is consistent with the size of the fields found at Pulltrouser Swamp (Turner and Harrison 1981).

The large canals appear to occupy less than 10 percent of the area devoted to the canal network, and the platforms occupy about 60 percent of the *bajo* surface (Gliessman, unpublished data). This figure is higher than Denevan's (1979) calculations based on photo interpretation, but within the range found for other Maya fields (Turner and Harrison 1981). The irregularity of platform and canal sizes are also consistent with other Maya systems. The flatness of the *bajo* floor almost precludes drainage *per se* as the purpose of the system. Indeed, SARH's attempts to dig drainage canals to facilitate harvesting have met with little success.

The striking similarities between the results of analysis of the surface layers and the dark grey layer on the platforms lends more credibility to our argument that the original *bajo* surface was used for mound construction. Acidic pH, considerable residual organic matter, higher

Table 5:5 Rice growth and development on platforms and canals in the Bajo of Morocoy. Planted 20 May 1980, measured 6 September 1980.

| | Stems/$m^2$ | | Ave. Plant Ht. (cm) | | Biomass/$m^2$ gms dry wt | |
|---|---|---|---|---|---|---|
| Site 1 | | | | | | |
| Platform | 346.7 | 124.1 | 21.4 | 4.7 | 103.5 | 27.0 |
| Canal | 2457.3 | 201.4 | 68.1 | 18.6* | 866.1 | 150.5 |
| Site 2 | | | | | | |
| Platform | 829.3 | 478.7 | 23.4 | 4.6 | 122.5 | 61.6 |
| Canal | 1466.7 | 264.0 | 53.5 | 14.8 | 676.0 | 61.4 |

* Plants flowering in the canal at Site 1

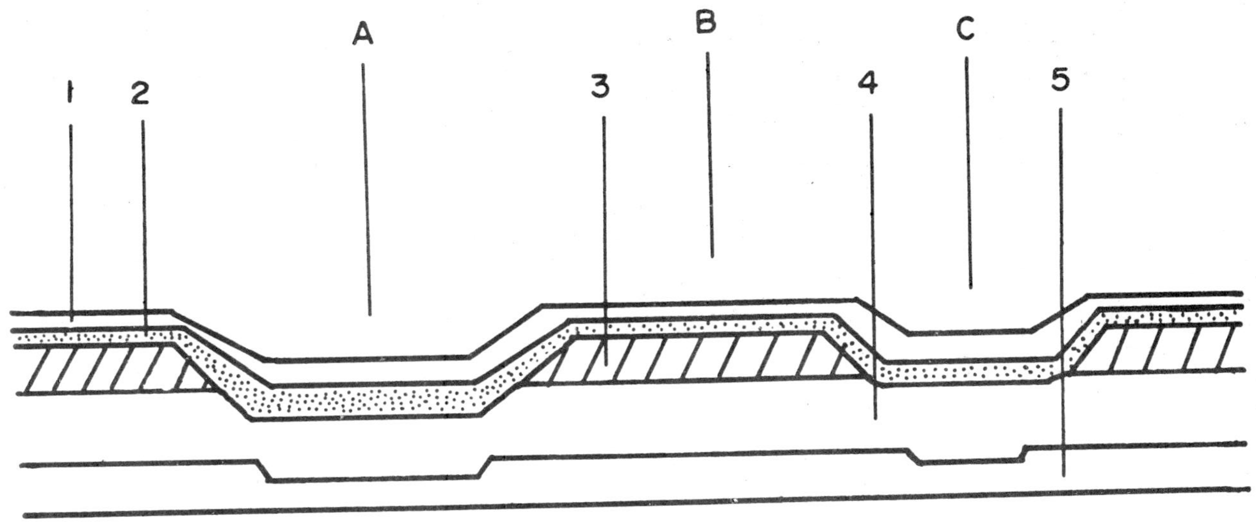


<u>Figure 5:2</u> Diagrammatic representation of the raised fields in the Bajo of Morocoy (not to scale). A = main canal, 8-10m wide; B = platform, 15-20m wide; C = secondary canal, 6-7m wide; 1 = dark surface forest soil; 2 = speckled brown recent fill; 3 = dark grey raised platform; 4 = light grey clay; 5 = white decomposed limestone bedrock (<u>sascab</u>)

levels of phosphate, and lower levels of calcium are all differences that could have been established by mounding up the original topsoil, and then maintained by a combination of periodic canal clearing and agricultural practices which allowed for constant organic matter inputs. Higher total nitrogen and nitrates in the canals may be more a response to fertilizers applied to the contemporary rice crop which has washed out into the depressions, rather than any reflection of past use. Periodic excavation of the canals and consequent contact with the calcium rich limestone below could account for the higher levels of that element there, similarly for magnesium. Constant high humidity below the surface layer would have aided greatly in maintaining higher organic matter levels, as well as associated acidity. Further analysis of certain pedological features of the soil, especially clay mineral analysis and soil texture, might provide further verification that the fields were used in the past.

The high moisture levels in the soil, even in the dry season, leads us to believe that the soils of the *bajos* probably never dry out to any significant degree, especially when still covered by forest. This would have greatly limited agricultural use of such wetlands without constructing raised fields. Encountering water at 1.5m in one of our excavations helps explain how this moisture is maintained. Only the upper 45 cm, directly affected by high solar radiation, showed appreciable dryness, and this would be greatly reduced with the continual vegetative cover. The argument that a type of "gilgai" activity might have been responsible for the formation of the platforms (Puleston 1978) would not seem plausible with moisture levels as high as we observed. The response of the rice crop aided us in more clearly outlining the topographic diversity which undoubtedly would have been much clearer if the area had not been altered so greatly by heavy machinery during the two years previous to our study. Careful clearing of adjacent sections of the forested *bajo* could provide information on this point.

We have only limited evidence on the types of cultivation practices associated with these relic *bajo* systems. Evidence from Pulltrouser Swamp and Albion Island in Belize indicates that maize and cotton may have been component crop species (e.g. Turner and Harrison 1981). Otherwise, by studying present day local use of wetlands in southeastern Mexico (Gliessman 1979; West *et al*. 1969), perhaps derived in part from practices used during Maya times, certain implications can be made as to some of the past uses.

A system for planting wetlands as the floodwaters recede in March (called the *Marceno*) has been described for Tabasco (Orozco-Segovia and Gliessman 1979), alternating with the planting of adjacent higher land at wetter times of the year. The fact that terminology for the varieties and activities has Maya roots has important implications. Maize and beans form the basis for this cropping system, just as they may have in the past. Such practices are well suited to periodically flooded lands and can allow for continual crop production throughout the year. Initial planting of the raised platforms probably took place at the outset of the rains in early June when conditions in the canals would have become too wet for planting. Harvest would have taken place in late September, with moisture levels in normal years beginning to fall in October. Planting of the depressions then would have been possible in late November for harvest during the early dry season. If the season turned out to be very wet, a second crop could have been planted on the platforms instead, being ready to harvest in late January or early February. Following the two wet season crops, a dry season planting could also be made in the canals as late as March by using short cycle varieties of cultivars still used today in much of rural southern Mexico (Orozco-Segovia and Gliessman 1979). It is unknown with which frequency the canals may have been cleaned for mounding up of the platforms, something of course which would have altered this cycle. The probability of two crops and the possibility of three crops can be postulated. This intensity of output could have been maintained ecologically by frequent "mucking" of the platforms, applying soil from the canals, or through a diversity of cropping practices still observable in Mexico based on a combination of such concepts as multiple cropping, high organic matter inputs, and biological control (Gliessman 1979; 1981; Gliessman and Amador 1980; Gliessman _et al_ 1981). Field tests are currently underway in Tabasco to attempt to reconstruct and manage such an agroecosystem (Amador _et al_. 1980). The field to canal system postulated here differs from that postulated for the fields at Pulltrouser Swamp where surface water is much deeper and sustained throughout the year.

The frequency of cultivation on the fields is uncertain, although it is assumed to be high, given the labour costs (Turner and Harrison 1981; Turner Chapter 2) of construction and the ecological contexts that are suitable for continuous-type cropping. The area is reported to contain a high density of structures, suggesting a once large populace which would have placed high demands on production systems (Harrison 1981). We can only speculate as to production for trade and other factors, but suggest that settlement studies will indicate a large populace was present in the area during, at least, the late Classic Period (e.g. Harrison 1981).

The eastern and southern uplands surrounding the large _bajo_ complex show high densities of structures almost everywhere the land is cleared and, of course, Nicolas Bravo and Tzibanche are ancient sites located just west and north of the _bajo_, respectively. In addition, we observed a large number of structures on a natural island near the southern edge of the _bajo_. And, surprisingly, we found a series of small stone-rubble structures (house-sites?) situated on raised-field-like platforms about 1.0 km north of our study area, closer to the interior of the bajo. To our knowledge, these are the first structures reported in a bajo per se (as opposed to the edge or on natural islands) in the Maya area. We had no archaeological permit and so were limited to surface observations. The structures had been levelled by tractor and chains, but prepared stone, ceramics and lithics were clearly visible on the surface, associated with elevated platforms. Sample ceramics were photographed and tentatively identified by Robert Fry as most probably Late Classic (A.D. 600-850) in origin, but possibly some Late Preclassic (800 B.C. - A.D. 150) specimens were present. These dates coincide with those found at Pulltrouser Swamp. The use of the platform-situated structures is not known, but it should be recognised that Bajo Morocoy is large, much larger than any other raised-field or channelised-field zone yet studied, and that the fields found there are situated at almost 9 km round trip distance to the nearest wetland margin. It may be that other large bajo complexes in Quintana Roo will contain some structures as well because of the distances to other land involved.

## Conclusions

The study of the Bajo Morocoy near Nicolas Bravo has provided the necessary ground survey evidence that the ancient Maya used raised fields to cultivate low-lying wetlands (bajos). This agroecosystem technology was apparently not so much to eliminate water from the system, but rather more a means of ensuring the presence of suitable planting surfaces, on one hand during the peak of the wet season (the tops of the platforms), and perhaps by taking advantage of water retained in the system during the dry season (the depressions). These procedures could have allowed for cultivation throughout the year. The few larger canals probably could have served to eliminate water from the bajo during the infrequent hurricanes sometimes occurring in the region, or acted as canoe routes to the production zone. The secondary canals appear to be more a remnant of the platform building processes, and only incidentally served for water control. The canals may never have contained water throughout the year, unless the environment was much different in the past or the Maya regulated water loss during the dry season. Regardless, the circumstance seems to differ from the year-round canal water postulated for Pulltrouser Swamp.

This study correlates well with aerial surveys and remote-sensing work demonstrating a very widespread use of wetland cultivation in prehistoric times (Scarborough Chapter 7). Our reassessment of the agricultural basis of the Maya civilization continues, especially in relation to sociopolitical structure, population densities, cropping systems and social organisation, and the environmental consequences of such intensive land use in association with the collapse of the Classic Maya civilization.

Acknowledgements

The authors are deeply indebted to the following staff members of the SARH, without whose assistance this study would not have been possible: Ing. Mario E. Villanueva M., Ing. Juan de Jesus Abarca P., Ing. Eduardo Franco V., Ing. Manuel Encalada F., Ing. Carlos Cervera S., and Ing. Mercadeo. We are most grateful to Jose W. Chan Puc for all of the aid, both direct and indirect, that he provided. Invaluable help was provided in the field by Silas Romero B., Arturo and Fernando Aguilar, and the students in the spring 1980 graduate Agroecology course (CSAT). Financial assistance from CSAT is acknowledged.

Bibliography

ADAMS, R.E.W. 1980, Swamps, canals and the location of ancient Maya cities, Antiquity 54, 206-214.

ADAMS, R.E.W., W.E. BROWN, Jr. and T.P. CULBERT 1981, Radar mapping, archaeology and ancient Maya land use, Science 213, 1457-1463.

AMADOR, M.F., S.R. GLIESSMAN, R. CORONADO,y.F. MIRAFUENTES 1982, El uso de bancales como alternativa en los suelos mal drenados, Agricultura Tropical (CSAT-Mexico), (in press).

DENEVAN, W.M. 1979, Raised field measurement comparisons, Summary of paper presented at XLIII Congress of Americanists, August, 1979, Vancouver, B.C., Canada.

GLIESSMAN, S.R. 1979, Some ecological relationships of traditional agroecosystems in the lowland humid tropics of Southern Mexico, Presented at the XLIII Congress of Americanists, Vancouver, B.C., Canada. Memoirs in preparation.

GLIESSMAN, S.R. 1981, Aspects of nitrogen distribution in several traditional agroecosystems in the humid tropical lowlands of southeastern Mexico, Ecological Bulletins (Sweden), (in press).

GLIESSMAN, S.R. and M.F. AMADOR 1980, Ecological aspects of production in traditonal agroecosystems in the humid lowland tropics of Mexico, In, Tropical Ecology and Development, edited by J.I. Furtado, Kuala Lumpur, 601-608.

GLIESSMAN, S.R., R. ESPINOSA E. and M. AMADOR A. 1981, The ecological basis for the application of traditional agricultural technology in the management of tropical agroecosytems, Agro-ecosystems 7, 173-185.

HARRISON, P.D. 1977, The rise of the *bajos* and the fall of the Maya, *In, Social process in Maya Prehistory: Studies in Memory of Sir Eric Thompson*, edited by N. Hammond, Academic Press, London.

HARRISON, P.D. 1978, *Bajos* revisited: visual evidence for one system of agriculture, *In, Prehispanic Maya Agriculture*, edited by P.D. Harrison and B.L. Turner II, University of New Mexico Press, Albuquerque, 247-254.

HARRISON, P.D. 1981, Some aspects of preconquest settlement in Southern Quintana Roo, Mexico, *In, Lowland Maya Settlement Patterns*, edited by W. Ashmore, University of New Mexico Press.

HARRISON, P.D. and B.L. TURNER II (editors) 1978, *Prehispanic Maya Agriculture*, University of New Mexico Press, Albuquerque, 414pp.

JENNY, H. 1980, *The Soil Resource: Origin and Behaviour*, Ecological Studies 37, Springer-Verlag, New York, 15-17.

LAMBERT, J.D.H. 1979, Raised field agriculture adjacent to the New River Lagoon, Lamanai, Belize: Paper presented at the XLIII International Congress of Americanists, Vancouver, British Columbia, Canada.

LOPEZ, J.H., J. LANDEROS G., F. LOPEZ M., J.H. PECH L., F. GALICIA S. and I. RENTERIA A. 1977, Estudio agrologico semidetallado de la Cuenca del Arroyo Ucum, Quintana Roo, *Serie Estudios*, Publ. 11, (Servicio de Agricultura y Recursos Hidraulicos), Mexico, 81pp.

OROZCO-SEGOVIA, D. and S.R. GLIESSMAN 1979, The *Marceno* in flood-prone regions of Tabasco, Mexico, Presented at the XLIII Congress of Americanists, Vancouver, B.C., Canada, Memoirs in preparation.

PENNINGTON, T.D. and J. SARUKHAN 1968, *Arboles Tropicales de Mexico*, I.N.I.F., FAO, Mexico.

PULESTON, D.E. 1977, Experiments in prehistoric raised field agriculture: learning from the past, *Journal Belizean Affairs 5*, 36-43.

PULESTON, D.E. 1978, Terracing, raised fields and tree cropping in the Maya lowlands: a new perspective in the geography of power, *In, Prehispanic Maya Agriculture*, edited by P.D. Harrison and B.L. Turner II, University of New Mexico Press, Albuquerque, 225-246.

SANDERS, W.T. 1979, The jolly green giant in tenth century Yucatan, or fact and fancy in Classic Maya agriculture, *Reviews in Anthropology* 6, 493-506.

SCARBOROUGH, V.L. 1980, The Settlement System in a Late Preclassic Maya Community: Cerros, Northern Belize, Ph.D. dissertation, Department of Anthropology, Southern Methodist University, Texas, 472pp.

SECRETARIA DE AGRICULTURE Y RECURSOS HIDRAULICOS (SARH) 1977, Mosaico de la Zona de Ucum, Quintana Roo, Fecha de vuelo Marzo 1977, Direccion de Estudios Especificos, SARH, Mexico, D.F.

SECRETARIA DE AGRICULTURE Y RECURSOS HIDRAULICOS (SARH) 1980, Diario de avances del cultivo de arroz, Laguna Om, Unpublished records, Chetumal, Quintana Roo, Mexico.

SIEMENS, A.H. 1978, Karst and the prehispanic Maya in the southern lowlands, In, Prehispanic Maya Agriculture, edited by P.D. Harrison and B.L. Turner II, University of New Mexico Press, Albuquerque, 117-144.

SIEMENS, A.H. 1982, Prehispanic agricultural use of wetlands of Northern Belize, In, Maya Subsistence: Studies in Memory of Dennis E. Puleston, edited by Kent V. Flannery, Academic Press, London, 205-225.

SIEMENS, A.H. and D.E. PULESTON 1972, Ridged fields and associated features in southern Campeche: new perspectives on the lowland Maya, American Antiquity 37, 228-239.

TURNER, B.L. II 1974, Prehistoric intensive agriculture in the Mayan lowlands, Science 185, 118-124.

TURNER, B.L. II 1978, Ancient agricultural land use in the central Maya lowlands, In, Prehispanic Maya Agriculture, edited by P.D. Harrison and B.L. Turner II, University of New Mexico Press, Albuquerque, 163-184.

TURNER, B.L. II 1979, Prehispanic terracing in the Central Maya Lowlands: Problems of agricultural intensitification, In, Maya Agriculture and Ethnohistory, edited by N. Hamond and G.R. Willey, University of Texas Press, Austin, 103-115.

TURNER, B.L. II and P.D. HARRISON 1979, Comment on W.T. Sanders Review of Prehispanic Maya Agriculture, Review in Anthropology 6, 544-555.

TURNER, B.L. II and P.D. HARRISON 1981, Prehistoric raised-field agriculture in the Maya lowlands, Science 213, 399-405.

TURNER, B.L. II and P.D. HARRISON 1983, Maya Raised-Field Agriculture and Settlement at Pulltrouser Swamp, Northern Belize, University of Texas Press, Austin.

WEST, R.C., N.P. PSUTY and B.G. THOM 1969, The Tabascan Lowlands of Southeastern Mexico, Louisiana State University Press, Baton Rouge, 193pp.

Chapter 6

# ANCIENT MAYA LAND USE AND POTENTIAL AGRICULTURAL PRODUCTIVITY AT LAMANAI, BELIZE

J.D.H. Lambert and J.T. Arnason

Abstract

The recent archaeological literature on Pre-Hispanic Maya agriculture indulges in much unfounded speculation on crop yields and the relative merits of intensive (sedentary) versus extensive (shifting) agriculture. In this paper a drained field complex on the New River near Lamanai, Belize, with evidence of Preclassic and Classic occupation was identified. Modern milpa agriculture was studied on 30,000 acres of forest to the west of Lamanai where evidence of ancient house mounds is abundant. The soil quality of the milpa and drained field sites and their crop yields were compared. It is proposed that drained field agriculture provided the major crops during the dry season and milpas during the wet season. It is also suggested that the Maya were probably well aware of the optimal land use methods for agriculture and would have used whatever method(s) were best suited to the topography and drainage conditions.

Resumen

La literatura arqueológica reciente sobre la agricultura maya pre-hispánica se permite el lujo de hacer conjeturas sin base acerca de rendimientos de cultivos y los méritos relativos de la agricultura intensiva (sedentaria) contra la agricultura extensiva (de traslado). Se identificó un complejo de campo drenado en el Rio Nuevo cerca de Lamanai, Belice, con evidencia de ocupación Pre-Clásica y Clásica. Se estudió la agricultura milpa moderna en 30.000 acres de selva al oeste de Lamanai donde existe abundante evidencia de antiguas lomas de casas. Se compararon la calidad del suelo de la milpa y de los sitios de campos drenados y sus rendimientos de cultivos. Se propone que la agricultura de campos drenados proporcionaba los cultivos principales durante la estación seca y las milpas durante la estación lluviosa. También se sugiere que los mayas probablemente estaban muy conscientes de los métodos óptimos del uso de la tierra para la agricultura y habrían usado el método que fuera más apropiado a las condiciones topográficas y de drenaje.

## Introduction

Hypotheses as to the types of agricultural systems practiced in Prehispanic Mesoamerica have ranged from extensive or milpa agriculture as the major form of subsistence in drier upland sites (Morley 1946) to intensive, upland terrace and raised field agriculture in wetland sites (Harrison and Turner 1978).

Today wherever forest or bush are removed evidence of past occupancy and/or agricultural activity are invariably found. New methods of air survey, such as Synthetic Aperture Radar (SAR) have identified large areas in the forested lowlands of Mexico, Guatemala and Belize that are covered with terraces and raised fields (Adams, 1982). In a number of cases, areas, such as bajos, which would today be classified as of low productivity, were turned by the Maya into productive agricultural land. Such evidence might be used to suggest that the Maya were aware of the productivity of different landforms so that they could have utilized both extensive and intensive agricultural practices.

It has been suggested that raised field agriculture would be more efficient than slash and burn in that it avoided the inherent inefficiencies of the fallow cycle (Puleston 1977). However, burning and fallow have obvious benefits. They aid organic matter and nutrient returns to the soils, and control weeds and various pathogen populations. Extensive agriculture as defined by Harrison (1978), involving a milpa system in which intercropping and multicropping were practiced would, we suggest permit increased species diversity, help control weeds and probably allow for longer use without necessarily reducing the fallow period. Besides, the fallow period in such cropping systems could have been practiced as a crop disease protection tactic, rather than for strictly agronomic purposes as suggested by Glass and Thurston (1978).

Evidence is presented here for Prehispanic seasonal use of the New River floodplain near the Maya site of Lamanai in Belize. We also present data on modern day extensive (milpa) agriculture at Indian Church (Lamanai) to support our theory that in specific areas the Maya could have used both intensive and extensive systems to supply their basic nutritional requirements. To have developed as successful a culture as appears evident the Maya probably made optimal use of all available landforms and prevailing environmental conditions.

## Study Area

The northern part of Belize is physiographically a part of the Yucatan Peninsula and is a heavily karsted limestone plateau of Cretaceous age. Lamanai or Indian Church is located in the north east sector of this plateau on the west shore of the New River Lagoon at its entry into the New River. Both the river and lagoon are controlled at this point by a small fault scarp with Lamanai situated on the dip slope. The lagoon is maintained largely by groundwater flow and it has a relatively high sediment and mineral load. Erosion of the limestone scarp results in high rates of $CaCO_3$ deposition.

Lamanai is considered to have been in use for 2200 years or more (ca. 600 B.C. - A.D. 1675) with a population estimated at over 10,000 (Pendergast personal communication). In fact, the area was used continually by Maya farmers up until 1925 when an epidemic decimated the village of Indian Church. In the past decade farmers have again moved back into the region.

Immediately to the west of the New River Lagoon are 60,000 hectares of rolling land dominated by cohune palm and high bush forest, marsh forest (*bajo*) and herbaceous marsh and swamp (Figure 6:1). The region is underlain by a creamy pink, moderately hard, amorphous limestone. Soils in the area are those of the Yaxa group (Wright *et al*. 1959) being grey-brown clays. Yaxa shows the development of, a normal soil type on nearly level topography (high bush), a shallow gravelly soil on undulating land (high bush), a deep melanized soil under cohune palm forest and a mottled clay in areas of impeded drainage (*bajo* and swamps).

The vegetation is classified according to Wright *et al*. (1959) as, broadleaf deciduous forest and is included in the lowland, dry tropical zone. Five major forest associations have been identified and their species composition and soil characteristics documented (Lambert and Arnason 1978).

## Soils

Data for the chemical and physical characteristics of the most suitable sites for agriculture (Wright *et al*. 1959; Lambert and Arnason 1979) are presented in Table 6:1. The cohune and high bush forest soils reflect similar characteristics. The only major differences occur in exchangeable calcium and sodium and cation exchange capacity (CEC). The three values are indicative of the limestone bedrock that underlies the region.

The physical characteristics of the soil have a major effect on the degree of leaching and nutrient availability. The high clay and organic matter content of the soil guarantees an abundance of clay-humus micelles, i.e. sites of ion exchange. This situation plus the base rich condition indicates a fertile soil of good agricultural potential. Additional nutrients available for short term agricultural activity are stored in the cohune and high bush forests in the standing biomass. Such reserves are released on burning and a certain percent is made available for crop production.

A slash and burn/milpa type of agriculture, from our studies over a four year period, does not appear to reduce soil fertility. Reductions in crop yields are primarily related to increased weed growth, which is in itself an indicator of sufficient nutrients (Lambert and Arnason 1980).

An initial drop in exchangeable cation levels occurs at the start of the rainy season, in May, and indicates the initiation of plant growth as well as a short period of leaching. The high montmorillonite clay content of the

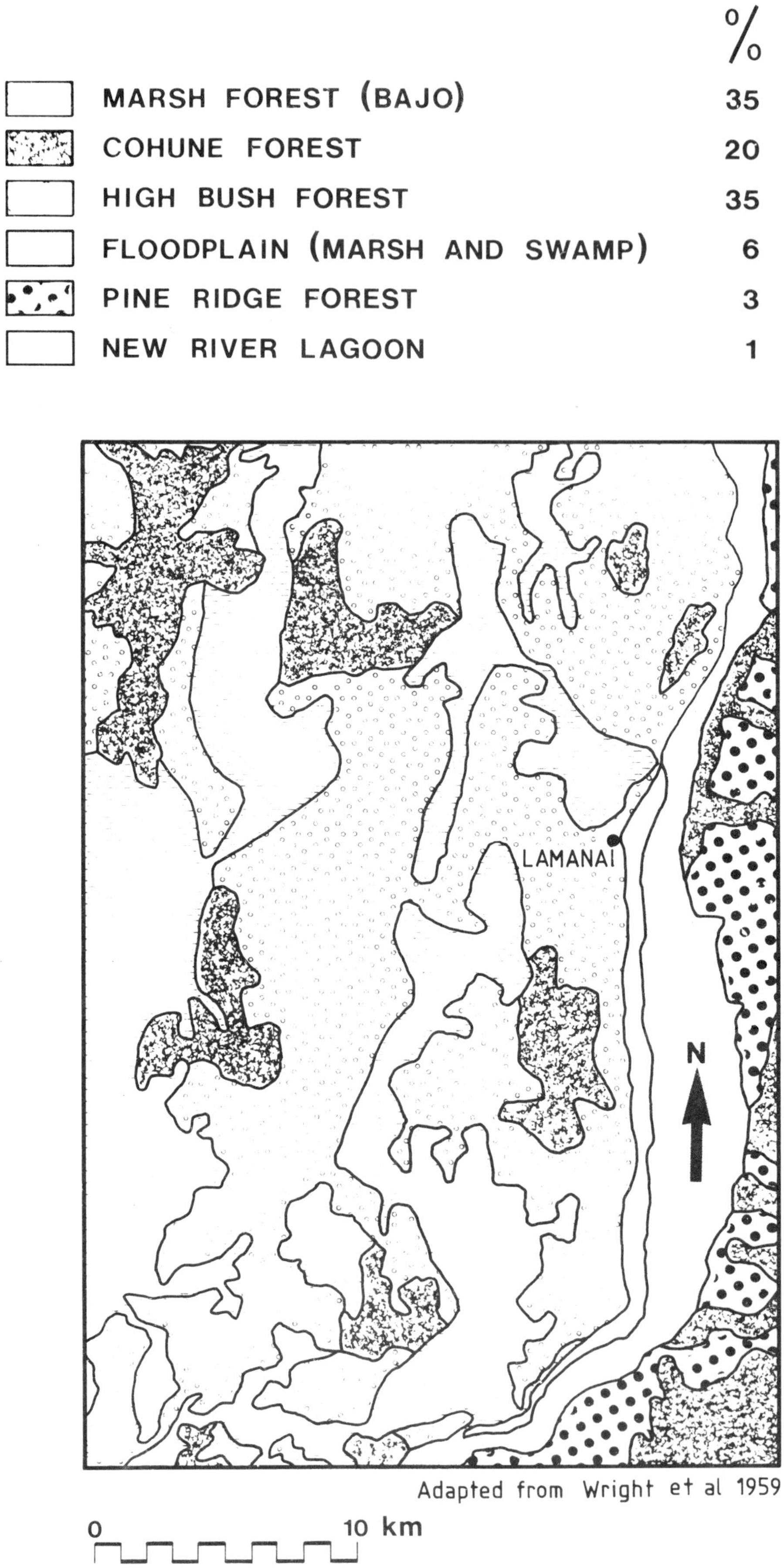


Figure 6:1 Vegetation map outlining major communities and their recent cover at Lamanai, Belize

| | Cohune Palm Forest | High Bush Forest | Drained Field | Annual Sediment Deposition |
|---|---|---|---|---|
| Exchangeable Cations ppm (1 g/100 ml) | | | | |
| Ca | 145.0 | 260.0 | 72.0 | 662.0 |
| Mg | 6.0 | 7.7 | 19.0 | 40.0 |
| Na | 2.0 | 0.6 | 6.7 | 3.0 |
| K | 9.0 | 7.2 | 1.5 | 5.0 |
| Available P | 0.32 | 0.32 | 0.2 | 1.6 |
| Micronutrients | | | | |
| Fe | 0.5 | 0.14 | 1.6 | 0.9 |
| Al | 1.3 | 1.5 | 6.0 | 5.0 |
| Mn | 0.1 | 0.7 | 0.3 | 0.6 |
| Zn | 0.04 | 0.04 | 0.02 | 0.1 |
| Cation Exchange Capacity | 125 | 210 | 181 | 275 |
| % Nitrogen | 0.31 | 0.46 | 0.6 | 1.7 |
| % Organic Matter | 14.1 | 20.2 | 32.0 | 59.0 |
| pH | 6.9 | 7.0 | 6.2 | 7.6 |
| C/N | 20:1 | 19:1 | 15:1 | 11:1 |
| Sand % | 43 | 28 | 19 | 52 |
| Silt % | 15 | 30 | 6 | 29 |
| Clay % | 42 | 42 | 75 | 19 |

Table 6:1 Summary of soil chemical and physical analyses of the A horizon in forest and floodplain soils, Indian Church, Belize.

soil however, would inhibit prolonged leaching due to its swelling properties when wet.

Percent moisture increases sharply at the advent of the rainy season. However, maximum levels are not reached until July. The start of the dry season in January results in a steady decline in soil moisture.

Soil pH values (averaged for the twelve month period) indicate the soil is neutral. With the start of the rainy season bases are removed by leaching giving rise to a slightly acid condition which in turn is responsible for the availability of micronutrients (Fe, Al, Zn, Mn, etc.). During the dry season the soil pH is slightly alkaline and micronutrients are unavailable.

In a recent paper (Lambert *et al*. 1983) we have discussed the agricultural potential of the drained field soils which represents a very different situation from the cohune and high bush forests. The annual deposition of nutrient rich sediments, followed by a dry season during which mineralization occurs rapidly, guarantees an abundance of nutrients every year. The high montmorillonite clay fraction would pick up available nutrients and in its swollen state reduce leaching and maintain a high soil moisture content during the dry season. Carbonates would be leached through the profile and accumulated in the gypsum layer. The annual flooding of the drained fields and deposition of sediments would inhibit or at least drastically reduce the ability of weeds to establish and survive.

Underlying the drained fields is a layer (at least 2m thick) dominated by gypsum with some calcite.

## Discussion

In areas of impeded drainage (*bajo* and swamp) west of the New River Lagoon we have found no evidence, to date, of raised or channelized fields that would indicate wetland cultivation by the ancient Maya as described by Turner and Harrison (1981) at Pulltrouser Swamp 50 km to the north. The only evidence we have of possible wetland cultivation is based on the presence of canals on the east shore of the New River Lagoon and in the New River flood-plain. The general undulating character of the topography west of the Lagoon and the dominance of high bush and cohune forest with potentially suitable soil and our observation of numerous ancient house mounds leads us to suggest that the Maya at Lamanai could have supported themselves by a combination of both slash and burn/milpa and drained field agriculture.

In contrast, several recent papers (Harrison and Turner 1978; Deevey *et al*. 1979) support the suggestion that milpa agriculture while adequate in the 17th century would be inadequate for growing populations. Very low crop yields are cited. Perhaps the reason some present day milperos obtain low yields is because the land they cultivate is of poor or marginal quality. If yields are low it is probably due to the fact that the best land is used for export crops (coffee, sugar cane, hemp, bananas, etc.); pasture for large herds of cattle or, the productive members of the village are employed in sugar cane farming, lumbering or trucking.

Much of the published data (Cowgill 1961; Urrutia 1967; Dickson 1980) on crop yields gives no cultivar type and seems to imply that only one crop/milpa/year is or was planted and harvested. Early Spanish documents describing life in the Alta Verapez region of the Peten identify house gardens, large milpas (comaracas) and intercropping (corn and beans) with two plantings per year (Renia and Hill 1980) as we have observed in Belize.

Sanders (1973) reported that current corn production in the Peten was about 800 kg/ha, a yield we consider very poor and indicative of over used land (Arnason et al. 1982). In addition the above yield was further reduced to 528 kg/ha because, we quote "... of the more primitive varieties of maize being used in the Pre-Classic", which we find to be unfounded and patronizing. Willey and Smith (1973) suggest there is no evidence the Maya practiced plant selection to improve yields or resistance to disease. We find these assumptions to be unacceptable.

Every milpero we have talked to concerning farming techniques and seed types instinctively recognised the relationship between seeds and their adaptation to specific environmental conditions. For example, local corn varieties are preferred to the introduced hybrids because they are less susceptable to weevil attack while in storage. An analysis of local varieties and introduced hybrid corn shows that antifeedant phenolics present in the local seed have been bred out of the hybrids (Fortier et al. 1983). The success of local varieties is based on the abilities of milperos to select for and preserve the genetic heritage of the crops. The fact that over 100 species of plant were being cultivated at the time of European conquest attests to the success of Prehispanic culture in plant domestication and breeding (Brown and Finsterbusch, 1982). Flannery (1968) more realistically depicted the early Mesoamerican peoples as a "practiced and ingeneous team of lay botanists and not a ragged and scruffy band of nomads".

A factor seldom considered in any discussion of Maya agriculture is that a milpero keeps two milpas under cultivation all the time. Each dry season he cuts, burns and plants a new milpa, while at the same time planting in the reburned milpa he cut the year before. In this way he reduces his work load by half. While the milperos task is not easy it is not necessarily as difficult as the Frairs reported (Renia and Hill 1980).

While we suggest that the Maya must have continued the practice of slash and burn/milpa agriculture during the Classic period we see obvious advantages in exploiting the river floodplains as well. Besides corn and beans that could be grown in a bottomland environment there are other potentially useful crops that would have received special attention from early residents. There are several such plants in the Maya Lowlands that thrive under seasonal flooding (National Academy of Sciences, 1975). They include: cocoyam (*Xanthosoma sagittifolium*), cassava (*Manihot esculenta*), canna (*Canna edulis*) and cacao (*Theobroma cacao*).

To increase soil fertility the Maya could have periodically cleaned out the canals thereby adding organic and mineral material to the surface of the drained fields. Another approach the Maya could have taken, at least on the floodplain, was the application of pure gypsum. A thin layer spread on the surface moving slowly through the top soil would improve soil permeability and deposit additional quartz and montmorillonite clay thereby increasing cation retention. Gypsum is often applied to land with beneficial effects to crops in North America. The result is an increase in pH, and the availability of P, Mo, Ca and Mg and in the percent base saturation.

We can speculate on the integration of milpa and drained field agriculture on the basis of soil conditions on the floodplain and in our milpa research. If we assume that the most productive period of use of a milpa was two years and that the drained fields could be used continually because of the annual nutrient deposition then, based on present day yields of local crops we can determine the amount of land (milpa and drained field) necessary to feed the estimated population at Lamanai and the surrounding inhabitants (Table 6:2).

In the Indian Church and San Jose Succotz (Arnason *et al*. 1982) regions of Belize all milperos plant at least two corn crops (Cosecha and Yaxking) and one bean crop per year. The average yields in the Indian Church area based on our own milpa studies (Lambert and Arnason 1980) and data collected from twenty-one local milperos for two crops per year, is ca. 3100 kg/ha. Such yields are attained using local varieties of white, yellow and black corn and yields higher than 4000 kg/ha are common. The yield is one and a half times that when hybrid corn is grown. Bean yields from our milpa studies range from 354 kg/ha for Pinto beans, to 520 kg/ha for Red Kidney beans and to 720 kg/ha for Black-eyed beans. All three bean species are planted in November and harvested in February. Most milperos usually plant squash and/or melon with their corn and beans to inhibit weed growth.

The lower yields in the milpa for the Yaxking planting are readily explained. While seed germination and stalk elongation are not inhibited during the initial growth period, grain growth and maturation are restricted due to

| | Yield in kg/ha | |
|---|---|---|
| | Cosecha | Yaxking |
| Milpa | May-September | December-April |
| Corn (grain) | 1600 | 1000 |
| Beans (grain) | 500 | |
| | 2100 | 1000 |
| Land cultivated | | |
| 400 ha | 840,000 kg | 400,000 kg |
| Drained Field | | January-April |
| Corn | | 1400* |
| Beans and root crops | | 1000* |
| Land cultivated | | 2400 |
| 200 ha | | 480,00 kg |

Total yield from 600 ha - 1,720,000 kg/yr

* estimated yields based on vegetative growth.

Table 6:2 Crop yields from milpa and drained field systems, Indian Church, Belize

a limiting water supply during the dry season.

The period of maximum productivity for the drained fields would be from January to May when water levels are receding. While soil moisture levels in milpa land drops below 30% during the dry season, this is not the case in the drained fields. Because of the water-retaining ability of the gypsum, the high nutrient content of the montmorillonite subsurface layer and the loose friable structure of the organic surface horizon, crops would not experience drought-like conditions. Preliminary data on corn production suggests that we could expect excellent growth in the dry season. Specialised perennial crops such as cacao and cocyam could act as shade trees to vegetable crops planted as water levels dropped. To ensure maximum crop yields the Maya would have had to breed for shorter life cycle crops, probably of three months duration. Today local campasinos in Tabasco use a local three month corn on their chinampas (Gliessman personal communication) as do some milperos at Indian Church on the Yaxa soils.

Of the approximately 412 $km^2$ or 41200 ha identified as available for agriculture only 4000 ha would be on the floodplain and could have been used continuously without any fallow. The remaining 372 $km^2$ or 37200 ha would have been available for milpa/extensive agriculture with plots being used for two years followed by a fallow period. If only 400 ha of the available cohune and high bush forest were needed each year for crops large areas of forest would probably never have been cleared for agriculture. We would suggest that a resident population of 14,000 inhabitants, assuming average annual consumption of 240 kg/yr (Offner 1980; Dickson 1980), would have required no more than 6 $km^2$ or 600 ha per year to supply their basic nutritional needs. This amount of land would be less than 1.5% of the total available. With such extensive forest on good soil the local residents would have had no need to modify the swamplands as was the case in other regions.

By understanding the soil/water/plant growth relationships of the different landforms of the region the local Prehispanic inhabitants would have guaranteed themselves a steady food supply. Milpas could have produced abundant crops during the rainy season and the drained fields during the dry season. The selection of local crop varieties for different seasons and environmental conditions would have given rise to considerable genetic diversity. The results would have been at least partial resistance or tolerance to pest and environmental stress.

## Bibliography

ADAMS, R.E.W. 1982, Ancient Maya canals: grids and lattices in the Maya jungle, Archaeology 35, 28-35.

ARNASON, J.T., J.D.H. LAMBERT, J. GALE, and H. VERNON 1982, Decline of soil fertility due to intensification of land use by shifting agriculturalists in Belize, C.A, Agro-ecosystems 8, 27-37.

BROWN, L.R. and G.W. FINSTERBUSCH 1972, Man and His Environment: Food, Harper and Row, New York, 208pp.

COWGILL, U.M. 1961, Soil fertility and the ancient Maya, Connetticut Academy Arts and Sciences, Transactions 42, 1-56.

DEEVEY, E.S., D.S. RICE, P.M. RICE, H.H. VAUGHAN, M. BRENNER and M.S. FLANNERY 1979, Mayan urbanism: impact on a tropical karst environment, Science 206, 298-306.

DICKSON, D.B. 1980, Ancient agriculture and population at Tikal, Guatemala: an application of linear programming to the simulation of an archaeological problem, American Antiquity 45, 697-712.

FLANNERY, K.V. 1968, Archaeological systems theory in early Mesoamerica, In, Anthroplogical Archaeology of the Americas, edited by B.J. Meggars, Washington, D.C., 67-87.

FORTIER, G., J.T. ARNASON, J.D.H. LAMBERT, J. McNEILL, C. NOZZOLILLO and B.J.R. PHILOGENE 1983, Local and improved corn cultivars in small farm agriculture in Belize, C.A.: their taxonomy, productivity and resistance to Sitophilus zeamais, Phytoprotection, in press.

GLASS, B.H. and H.D. THURSTON 1978, Traditional and modern crop protection in perspective, Bioscience 28, 109-115.

HARRISON, P.D. 1978, So seeds shall grow: some introductory comments, In, Pre-Hispanic Maya Agriculture, edited by P.D. Harrison and B.L. Turner II, University of New Mexico Press, Albuquerque, 1-11.

HARRISON, P.D. and B.L. TURNER II, (editors) 1978, Pre-Hispanic Maya Agriculture, University of New Mexico Press, Albuquerque, 414pp.

LAMBERT, J.D.H. and J.T. ARNASON 1978, Distribution of vegetation on Maya ruins and its relationship to ancient land-use at Lamanai, Belize, Turrialba 28, 33-41.

LAMBERT, J.D.H. and J.T. ARNASON 1979, Indian Church Milpa Project, Belize: report on field studies and laboratory analyses, Report to Canadian International Development Agency, Ottawa.

LAMBERT, J.D.H. and J.T. ARNASON 1980, Nutrient levels in corn and competing weed species in a first year milpa, Indian Church, Belize, C.A, Plant and Soil 55, 415-427.

LAMBERT, J.D.H., A.H. SIEMENS and J.T. ARNASON 1983, Ancient Maya drained field agriculture: its possible application today in the New River floodplain, Belize, C.A, Submitted to Agro-ecosystems.

MORLEY, S.G. 1946, The Ancient Maya, Standford University Press, 507pp.

NATIONAL ACADEMY OF SCIENCES 1975, Underexploited Tropical Plants With Promising Economic Value, Washington, D.C., 188pp.

OFFNER, J.A. 1980, Archival reports of poor crop yields in the early post conquest Texcocan heartland and their implications for studies of Aztec period populations, American Antiquity 45, 848-856.

PULESTON, D.E. 1977, Experiments in Pre-Historic raised field agriculture: learning from the past, Journal Belizean Affairs 5, 36-43.

REINA, R.E. and R.M. HILL 1980, Lowland Maya subsistence: notes from ethnohistory and ethnography, American Antiquity 45, 74-79.

SANDERS, W.T. 1973, The cultural ecology of the Maya lowlands: a re-evaluation, In, The Classic Maya Collapse, edited by T.P. Culbert, University of New Mexico Press, Albuquerque, 325-365.

TURNER, B.L. II and P.D. HARRISON 1981, Prehistoric raised-field agriculture in the Maya lowlands, Science 213, 399-405.

URRUTIA, R.V.M. 1967, Corn production and soil fertility changes under shifting cultivation in Uaxactun, Guatemala, Unpublished M.Sc. Thesis, Department of Agriculture, University of Florida, Gainesville, 158pp.

WILLEY, G.R. and D.B. SHIMKIN 1973, The Maya collapse: a summary view, In, The Classic Maya Collapse, edited by T.P. Culbert, University of New Mexico Press, Albuquerque, 457-545

WRIGHT, A.C.S., D.H. ROMNEY, R.H. ARBUCKLE and V.E. VIAL 1959, Land Use in British Honduras, Colonial Research Pub, 24 H.M.S.O. London.

Chapter 7

# RAISED FIELD DETECTIONS AT CERROS, NORTHERN BELIZE

Vernon Scarborough

## Abstract

Late Preclassic raised fields have been mapped and excavated at Cerros, a Maya site on the northern Belizean coast at Lowry's Bight. Raised fields appear to have been designed either to: (1) control or manipulate flowing water through these agricultural plots or (2) to drain and render cultivatable internally drained depressions or *bajos*. At Cerros a 6m wide canal encloses raised fields in what appears to be a man-made catchment. These and other raised fields at the site have been studied on radar imagery. Overlays of the mapped raised fields and canals were compared with a side-looking airborne radar overlay of the site to see if it was possible to accurately locate and map raised fields using radar imagery. It is concluded that some of the larger canals are visible on the imagery. The possibility that the lineaments interpreted as canals and fields were in fact natural gilgai or structural lineaments is refuted because of the close relationship between the radar overlays and the site maps.

## Resumen

Se han ubicado y excavado campos elevados de fines del período preclásico en Cerros, un sitio maya en la costa norte de Belice en Lowry's Blight. Al parecer, los campos se diseñaron ya sea para controlar o manejar corrientes de agua o para drenar y permitir el cultivo de bajos drenados internamente. En Cerros un canal de 6m de ancho encierra campos elevados en lo que parece ser una captación artificial. Estos campos elevados y otros más en el sitio se han estudiado en imágenes de radar. Sobreimposiciones (overlays) de los campos elevados y canales trazados en el mapa se compararon con una sobreimposición de radar desde el aire de vista lateral del sitio para ver si era posible ubicar y trazar en forma precisa los campos elevados mediante imágenes de radar. Se concluye que algunos de los canales más grandes son visibles en las imágenes. Se descartó la posibilidad que los alineamientos interpretados como canales y campos fueran gilgai naturales o alineamientos estructurales, debido a la estrecha relación entre las sobreimposiciones de radar y los mapas del sitio.

## Introduction

Recent investigations in the Maya Lowlands have revealed the presence of raised field agriculture. Aerial photography has documented ridged fields in the Candelaria Basin of southern Campeche (Siemens and Puleston 1972) and raised fields in Quintana Roo, Mexico (Harrison 1977; Turner and Harrison 1978). Additional remote sensing work has identified raised fields along the Hondo and the New

Rivers of northern Belize (Siemens 1978, 1982), along the Belize River of central Belize (Kirke 1980), and in parts of the Peten, Guatemala (Adams 1980). Ground surveys have confirmed the extensiveness of raised field agriculture throughout northern Belize (Turner and Harrison 1981; Siemens 1982) and in the Peten, Guatemala (Adams *et al.* 1981). Field excavations have also been carried out at Albion Island on the Hondo River in northern Belize (Puleston 1977), at Pulltrouser Swamp, near the New River (Turner and Harrison 1981), and at Cerros, near the mouth of the New River (Freidel and Scarborough 1982; Scarborough in press), see Figure 7:1. These excavations provide demonstrable evidence of the methods of construction and function of the fields as well as the dates of their appearance in the Maya Lowlands.

## Types of Raised Fields in the Maya Lowlands

The raised fields at Cerros date to the Maya Late Preclassic period and excavations at Albion Island and Pulltrouser Swamp suggest comparable initiation dates, though Classic period utilization was intensive. The sizes of the plots range from 10 x 10m to 40 x 40m, although they are not necessarily square. Each is constructed from earth and fill material brought in from varying distances.

The fields excavated so far appear to have had one of two functions (Freidel n.d.). The first was to control or manipulate flowing drainage water on river flood plains and lagoonal margins such as at Albion Island or Pulltrouser Swamp. This type, although labour intensive, did not require a great deal of landscape modification as drainage was external and the system was naturally flushed by the river. The second type of raised field occurs in internally drained *bajos* in enclosed basins or surrounding lakes where drainage of planting surfaces is the main function. Much of the present aerial photographic coverage of Quintana Roo and northeastern Peten reveals fields in these settings.

Although examples of riverine fields have been excavated in northern Belize and elsewhere, few examples of still water fields and canalization have been excavated. The Cerros fields may be the exception. They are situated near the New River, but they differ from the Pulltrouser fields to the south. The Cerros fields are in a man-made catchment similar to a natural, internally drained *bajo*. This type of drainage system is entirely rain-fed, the rain water being channelled into a basin flanked by canals at the field loci.

Although the extensive *bajo* zones throughout the lowlands may have been reclaimed for agricultural purposes without a great deal of effort, the Cerros scheme represents a deliberate and costly modification of a formerly elevated and well drained setting. Quarry operations necessary for the construction of the

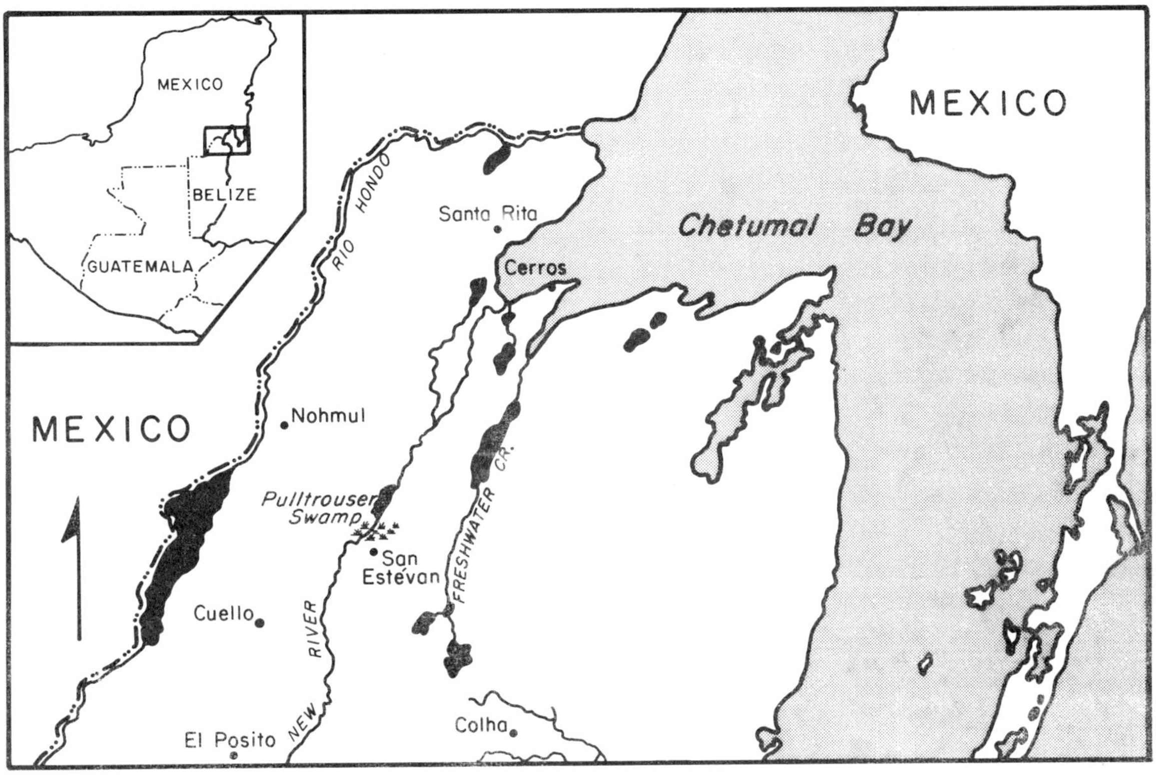


Figure 7:1 Map of Northern Belize

monumental architecture at Cerros are in part responsible for the man-made water catchment system (Scarborough in press).

## Cerros

Cerros, a Late Preclassic site is located on the coast at Lowry's Bight, northern Belize (Figure 7:1). The near absence of later construction at the site makes the community particulary significant for addressing the socioeconomic transition from Formative to Classic periods. Only a slight Classic and Postclassic occupation is in evidence at Cerros.

Systematic mapping and excavation of the site revealed a compact civic/residential core surrounded by a residential periphery (Figure 7:2). A major canal circles the core zone and encloses 95 mounded features within a 37 hectare area. Forty-one mounds have been excavated or horizontally exposed in the settlement zone. Over 1.5 $km^2$ of the 9.5 $km^2$ Bight have been surveyed with a theodolite.

Five microenvironments have been defined on Lowry's Bight using soils, drainage data and vegetation associations as the main mapping criteria. The zones defined are: (1) monte alto and huamil, or forest, (2) hulub bajo, (3) zacatal or grassland savanna, (4) thorn scrub savanna, and (5) a mangrove shoreline association. The monte alto and huamil grows on archaeological remains characterised by elevated, well drained soils which occur on or near plazas and mounds. Both hulub bajo and zacatal settings reflect the most low-lying, poorly drained soils at the site. At the margins of the zacatal the better drained soils associated with the thorn scrub savanna occur. Most of the identified raised fields appear in this zone. The mangrove association grows at the river mouth (Figure 7:3).

## Raised Fields at Cerros

The main concentration of raised fields rest in the south central portion of the core zone (Figure 7:4). These fields are low rectangular platforms, heavily eroded and much reduced from their original height. Each platform is circled by a narrow shallow channel. These canals widen and deepen to form periodic canal basins. The field plots are orientated to the cardinal directions and range in size from 14 x 28m to 22 x 40m. Although only five fields have been clearly defined in the area, another four may be present, but they are badly eroded.

The field plots and minor canals drain into the main canal at Cerros. The dimensions of the main canal are approximately 2m in depth, 6m in width and 1200m in length. Five trenches were dug across the main canal perpendicular to its long axis, and except for the southernmost trenches, the excavations were widely

Figure 7:2 Map of Cerros

Figure 7:3 Map of Cerros environs

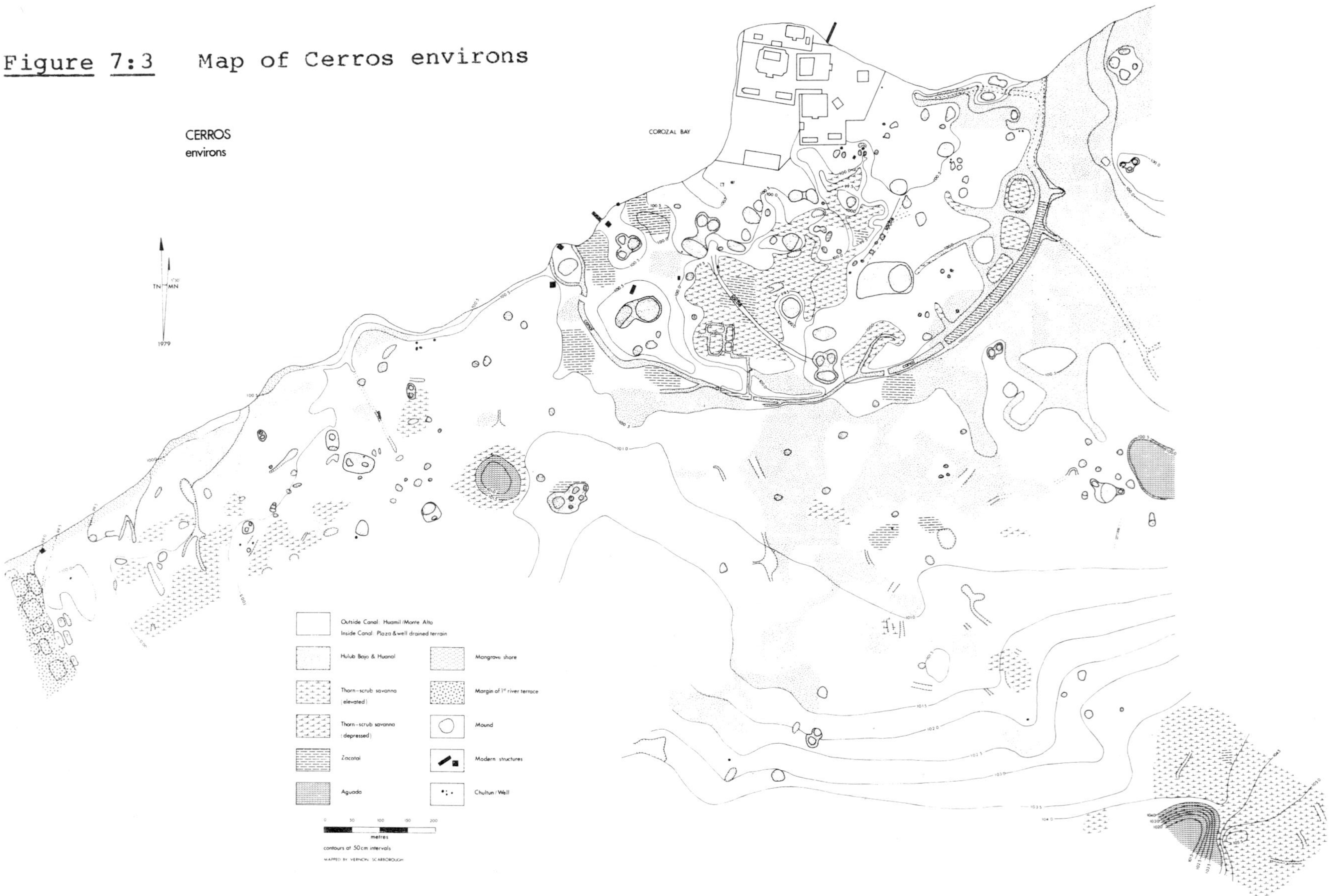

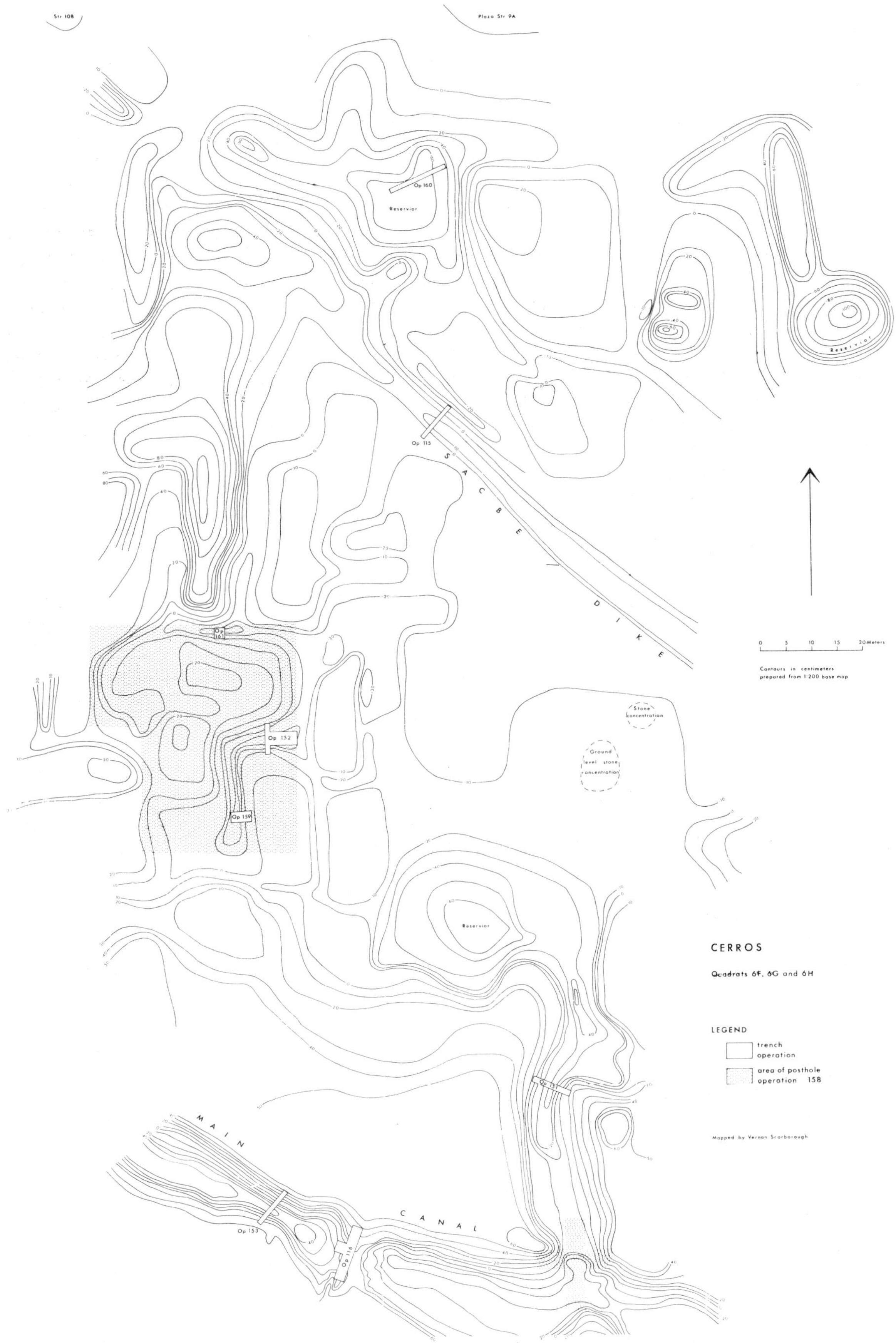


<u>Figure</u> <u>7:4</u> Contour map of the West Central core area. The known raised fields lie to the south and west of the <u>Sacbe</u> or causeway

separated. In addition, seven trenches and a systematic post-holing operation were excavated in the raised field complex.

Except for the field plots along the first river terrace at the south-west end of the shoreline survey traverse, most of the agricultural activity on the Bight is postulated to have been in immediate proximity to the main canal and within the 101.5m contour (datum assumed). Unfortunately, the vegetation and sedimentation in the *hulub bajo* setting has severely impaired the definition of these predicted features. However, the area to the east of the main canal has provided some suggestive contours in an area devoid of structures. It should be noted that the side-looking airborne radar (SLAR) readout (see below) suggests that these areas reflect a dendritic or lattice pattern (Figure 7:5 and 7:6). This pattern occurs all over the first river terraces of both the New River and Freshwater Creek and closely resembles the irregular field design of Lake Titicaca (Smith *et al.* 1968).

## Side-looking Airborne Radar

Side-looking airborne radar (SLAR) is a remote sensing technique which utilises the microwave portion of the electromagnetic spectrum to record information on the roughness and moisture characteristics of the surface. It has been developed and utilised by the Jet Propulsion Laboratory at Pasadena, California (Jensen *et al.* 1977; Adams *et al.* 1981). The technique has been employed because certain wavelengths can penetrate the dense semi-tropical and tropical forest to record ground surface information. The resulting imagery has revealed many patterned lineaments all over the Maya Lowlands which are believed to be ancient canals which retain water but are now infilled by fine grained sediments. Some of this imagery for the Cerros area has been analysed. It has been provided by Richard E.W. Adams in the form of transcribed overlays from the radar readout. Figures 7:5 and 7:6 were derived from this source.

In an attempt to assess the accuracy of the technique, the surveyed settlement patterns and canals at Cerros were used for comparison with the imagery. An overlay of the 1.5 $km^2$ surveyed area controlled by a 100m grid interval system was used. This has permitted an accurate assessment of the technique, though some problems with scaling have resulted. It should be noted that our survey and excavation programme was conducted before the imagery was available. Due to the small scale of the radar survey only the most pronounced archaeological features at Cerros are detectable.

Figure 7:5 reveals the extent of the side-looking airborne radar lineaments on Lowry's Bight and within the systematic survey zone. The main canal is depicted and the edge of the 2m high main plaza supporting the 6 ha. central precinct of the community is apparent. An area

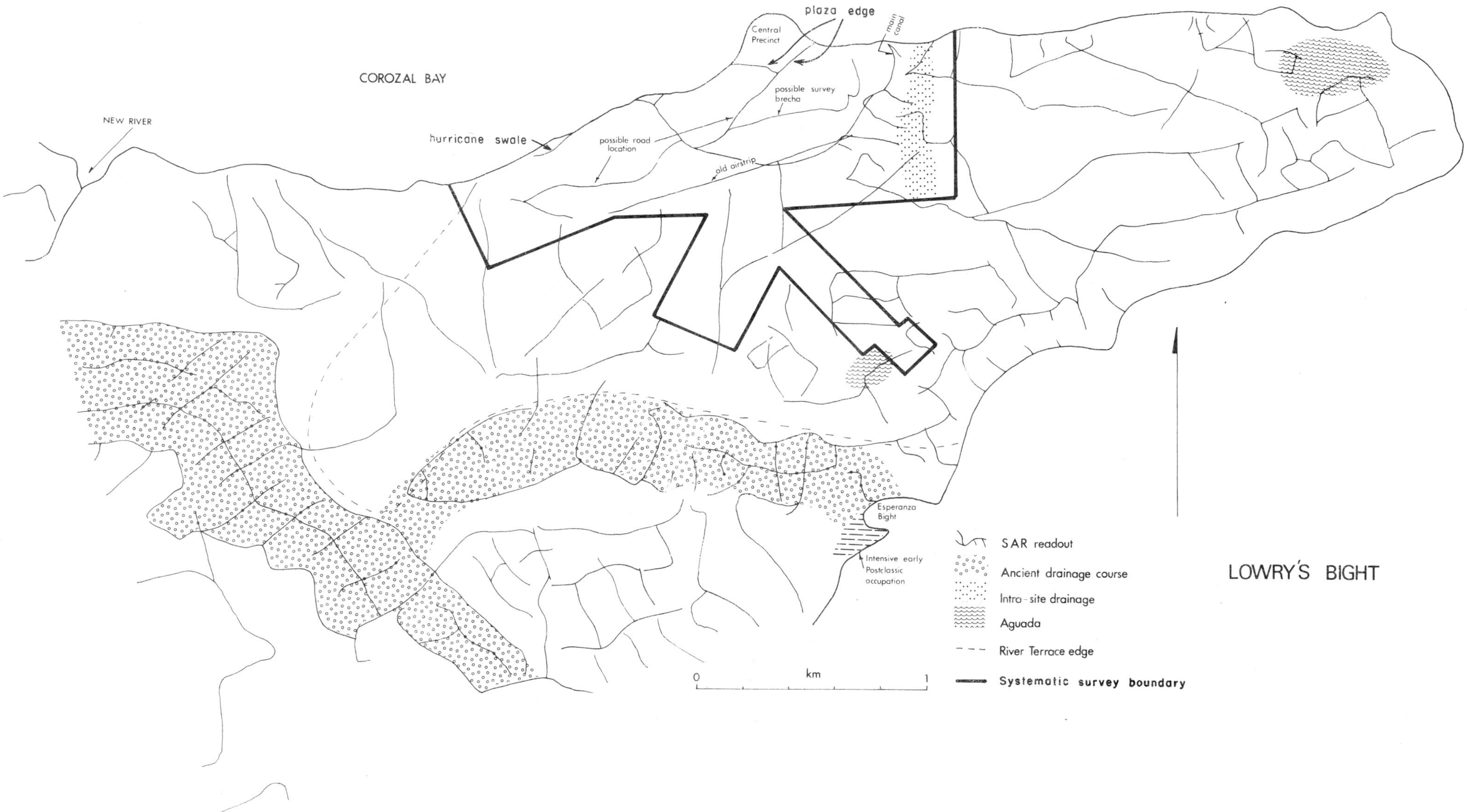


Figure 7:5 Map of Lowry's Bight (the SAR readout was provided by R.E.W. Adams)

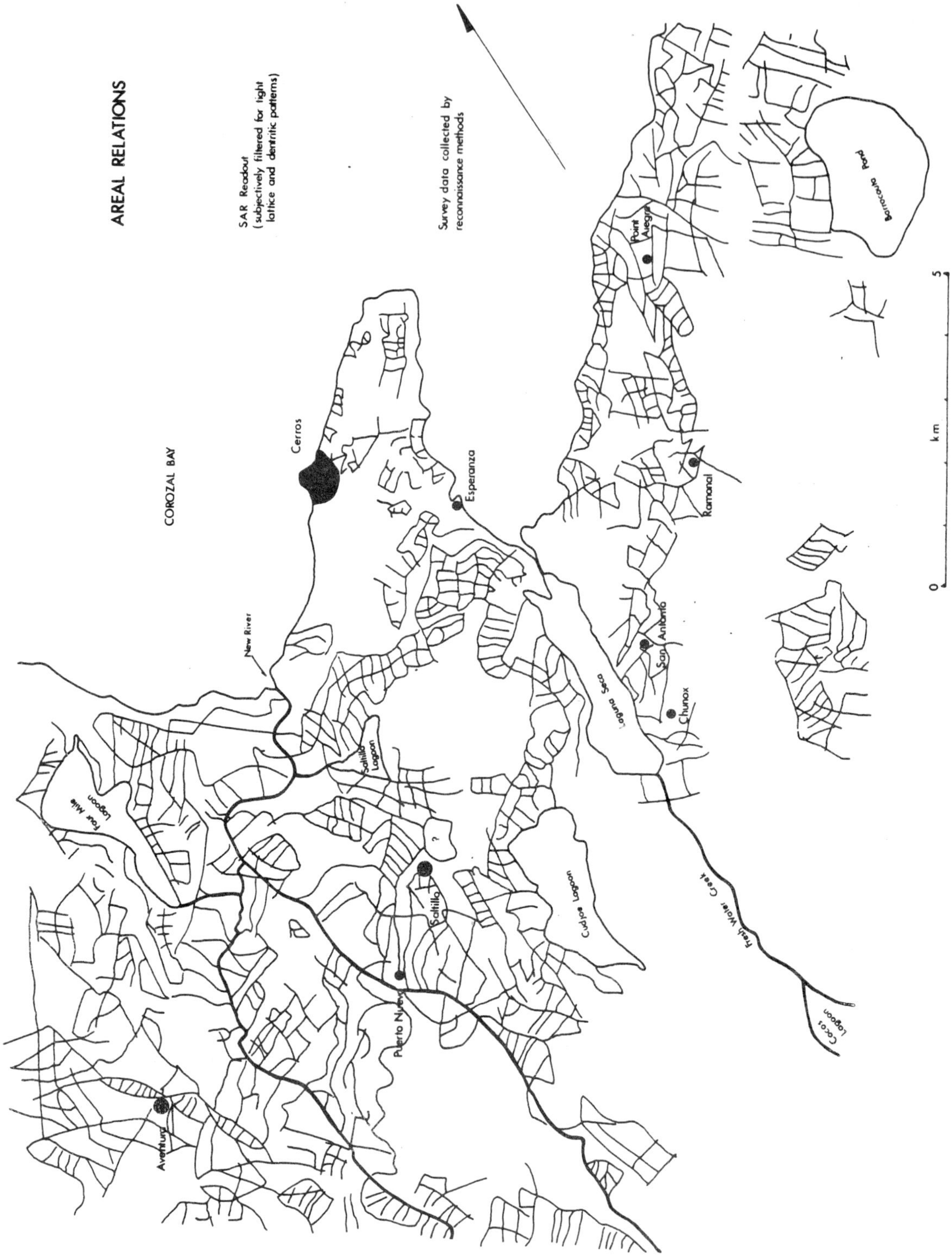


Figure 7:6 Map of Lowry's Bight and adjacent archaeological sites. The size of the black dot refers to the predicted size of those sites relative to Cerros (The SAR readout was provided by R.E.W. Adams)

immediately east of the canal associated with the *hulub bajo* setting also indicates a concentration of lineaments in a zone suspected of containing raised fields. Computation of overall lineament line length to identified lineament line length suggests that 20% of the side-looking airborne radar lineaments are drainage features at Cerros (Adams *et al*. 1981). Logging roads, an abandoned jungle airstrip, hurricane beach swales, and a possible well travelled survey *brecha* or footpath are also indicated.

Explanations of the lineaments as man-made features in the area of Cerros are confirmed by the close match between the position of the canals on the surveyed maps and on the imagery but, natural features could account for similar markings. The gilgai thesis which attributed such markings to vertic soil activity created by clays which shrink and swell in time with the seasonality of the climate has been laid to rest in northern Belize because the soils are too wet (Turner and Harrison 1981; Scarborough 1980) but structural faulting and jointing of the limestone bedrock may account for some of the patterning (see Goetz and Rowen 1981). Some geologists suggest that structural lineaments, joints and fracture traces will reveal a consistent range of orientations for an area rather than produce a bisecting pattern which is reminiscent of raised fields and canals (Gold 1980). Given that Tertiary faulting and folding have produced a series of near parallel ridges and troughs running the length of northern Belize though (McDonald 1979), a clear orientation of the patterns on the imagery should be in evidence if these lineaments are fracture traces. Visual inspection of the data does not suggest such an orientation. However, computer digitization of the radar imagery for portions of northern Belize is anticipated and this will provide orientational data as well as further line length ratios.

Figure 7:6 provides a subjectively filtered SLAR image overlay designed to enhance tight lattice and dendritic patterns in proximity to Cerros (compare Adams *et al*. 1981:Fig. 4). This particular area has not been carefully surveyed on the ground, though raised fields have been clearly identified in portions of the zone (Scarborough 1980; Siemens 1982). The area studied represents communities dating from the Late Preclassic or Early Classic periods (except for Esperanza which is an Early Postclassic site). The size of the black dot on Figure 7:6 represents the approximate size of the Prehispanic community. Although speculative, it is suggested from the evidence gathered to date that many of the lineaments define major field plots and canals which were constructed during the Late Preclassic and Early Classic periods.

## Conclusion

The occurrance of raised fields and canal systems have been confirmed in the Maya Lowlands. Excavated examples of these systems have revealed two functional types: still water canals generally associated with lake margins and flowing water canals associated with river floodplains. The Late Preclassic water system at Cerros indicates an example of an internally drained, still water agricultural system. Although the few demonstrable raised fields inside the community core zone at Cerros may be more symbolic and ornamental than agriculturally productive, their presence suggests the likelihood of intensive agriculture at an early date.

The side-looking airborne radar imagery appears to be a useful tool in the identification of some canal systems. The Cerros case study examined the imagery after the field survey was complete, so only the most apparent lineaments could be confirmed as canals. This test has provided one check on the utility of the technique, though additional ground survey is in order.

## Acknowledgements

I thank David Freidel, Robin Robertson, Susan Lewenstein, Beverly Mitchum, Eleanor King and Denise Seymour for their field assistance and R.E.W. Adams who generously provided the side-looking airborne radar readout overlay. This work was made possible through a research grant from the National Science Foundation to David Freidel at Southern Methodist University as well as a dissertation improvement grant from N.S.F. and a seed grant from the Institute for the Study of Earth and Man at Southern Methodist University. I am indebted to Harriot Topsey, Archaeological Commissioner of Belize and Jamie Awe as well as to the kind and diligent people of Chunox and Copperbank villages. Karim Sadr prepared the illustrations.

## Bibliography

ADAMS, R.E.W. 1980, Swamps, canals, and the locations of ancient Maya cities, Antiquity 54, (212), 206-214.

ADAMS, R.E.W., W.E. BROWN, Jr., and T.P. CULBERT 1981, Radar mapping, archaeology, and ancient Maya land use, Science 213, 1457-1463.

FREIDEL, D.A. and V. SCARBOROUGH 1982, Subsistence, trade and development of the coastal Maya, In, Maya Subsistence: Studies in Memory of Dennis E. Puleston, edited by K.V. Flannery, Academic Press, New York, 313-151

FREIDEL, D.A. n.d., Agricultural intensification and surplus: reflections on the Cerros field system, Paper presented at the 47th Annual Meeting of the Society of American Archaeology, Minneapolis, 1982.

GOETZ, A.F.H. and L.C. BROWN 1981, Geologic remote sensing, Science 211, (4484), 781-791.

GOLD, D.P. 1980, Lineaments and fracture traces, In, Remote Sensing in Geology, edited by B.S. Siegal and A.R. Gillespie, Wiley and Sons, 419-484.

HARRISON, P.D. 1977, The rise of the bajos and the fall of the Maya, In, Social Process in Maya Prehistory, edited by N. Hammond, Academic Press, New York, 470-509.

JENSEN, J., L.C. GRAHAM, J. PORCELLO and E.N. LEITH 1977, Side-looking Airborne Radar, Scientific American 237, 84-95.

KIRKE, C.M. 1980, Prehistoric agriculture in the Belize River Valley, World Archaeology 11, 281-286.

McDONALD, R.C. 1979, Preliminary report on the physical geography of Northern Belize, In, Cuello Project 1978 Interim Report, Archaeological Research Programme, Publication 1, New Brunswick, 79-87.

McDONALD, R.C. 1979, Tower karst geomorphology in Belize, Zeitscrift fur Geomorphologie, Supp. 32, 35-45.

PULESTON, D.E. 1977, The art and archaeology of hydraulic agriculture in the Maya Lowlands, In, Social Process in Maya Prehistory, edited by N. Hammond, Academic Press, New York, 449-467.

SCARBOROUGH, V.L. 1980, The Settlement System in the Late Preclassic Maya Community: Cerros, Northern Belize, Ph.D. dissertation, Southern Methodist University, University Microfilms, Ann Arbor, 472pp.

SCARBOROUGH, V.L. 1983, A Preclassic water system, American Antiquity 48, (4).

SIEMENS, A.H. 1978, Karst and the Prehispanic Maya in the Southern Lowlands, In, Prehistoric Maya Agriculture, edited by P.D. Harrison and B.L. Turner II, University of New Mexico Press, Albuquerque, 117-144.

SIEMENS, A.H. 1982, Prehistoric agricultural use of the wetlands of Northern Belize, In, Maya Subsistence: Studies in Memory of Dennis E. Puleston, edited by K.V. Flannery, Academic Press, New York, 205-222.

SIEMENS, A.H. and D.E. PULESTON 1972, Ridged fields and associated features in Southern Campeche, American Antiquity 37, 228-239.

SMITH, C.T., W.M. DENEVAN, and P. HAMILTON 1968, Ancient ridged fields in the region of Lake Titicaca, Geographical Journal 134, 353-367.

TURNER, B.L. II and P.D. HARRISON 1978, Implications from agriculture for Maya prehistory, In, *Prehispanic Maya Agriculture*, edited by P.D. Harrison and B.L. Turner II, University of New Mexico Press, Albuquerque, 337-374.

TURNER, B.L. II and P.D. HARRISON 1981, Prehistoric raised field agriculture in the Maya Lowlands, *Science* 213, 399-405.

Chapter 8

# EVOLUCION HISTORICA DEL REPERTORIO DE PLANTAS CULTIVADAS EN LAS CHINAMPAS DE LA CUENCA DE MEXICO

Mta. Teresa Rojas Rabiela


## Abstract

Agriculture in the chinampas is undoubtedly the best historically documented agricultural tradition of precolonial origin of Middle America. Furthermore, its practice in precise zones of the Basin of Mexico has been interrupted in five centuries and it is possible today to observe and record it. Documentation and this knowledge of a live tradition makes the reconstruction of the history of diverse aspects of this ancient agriculture possible. This paper deals basically with two of these aspects: the repertory of cultivated plants and the agricultural methods used. Changes which have taken place in both areas are discussed and there is an attempt to explain their relation to the ecological transformations which have occurred in the Basin during the period of study.

## Resumen

La agricultura en chinampas es sin duda la tradición agrícola de origen precolonial mejor documentada históricamente de Mesoamérica. A este hecho se suma el que su práctica en zonas precisas aunque más o menos reducidas de la Cuenca de México, no se ha interrumpido en cinco siglos y es posible observarla y registrarla hoy. La documentación y este conocimiento de la tradición viva, permiten la reconstrucción de la historia de diversos aspectos de esta antigua agricultura. En esta ponencia se tratan básicamente dos de ellos: el repertorio de plantas cultivadas y los métodos agrícolas. Se exponen los cambios ocurridos en uno y otro renglón y se intenta explicar su relación con las transformaciones ecológicas ocurridas en la Cuenca en el período de estudio.

## Introducción

La práctica ininterrumpida de la agricultura chinampera en las zonas de Xochimilco, Tláhuac, Itztapalapa y Mízquic, es de enorme interés actual e histórico. Actual, porque sus antiguos creadores y sus modernos innovadores, al mismo tiempo que han transmitido con vigor la centenaria tradición, sus complejos quehaceres y refinadas técnicas, han sabido adaptarla a las cambiantes condiciones ecológicas, sociales y culturales de la región. Histórico, porque su conocimiento es clave para entender el desarrollo cultural y social de la Cuenca de México - región de gran importancia en la historia del país - así como porque sirve para compararla con sistemas similares que existieron en toda América en tiempos precolombinos.

El caso de la agricultura chinampera, como bien lo hizo notar W. Denevan, es único en América puesto que es uno de los pocos sistemas agrícolas indígenas que no fueron abandonados con la conquista española, perviviendo hasta nuestros días. Sistemas emparentados, de camellones hechos en tierras con exceso de humedad que el hombre acondicionó con su trabajo para la agricultura, los hubo en época prehispánica desde Mesoamérica hasta el área andina. Ninguno, sin embargo, fue descrito ni se conoce con tanto detalle como el de la Cuenca de México (Zucchi y Denevan 1979:8-16).

Las chinampas mismas, el medio ambiente donde se desarrollaron, así como los modos de su agricultura, han despertado gran interés y curiosidad, más allá de los provocados por cualquier otra agricultura indígena de México. El resultado de ello es la producción de toda una gama de registros que va desde menciones breves y circunstanciales hasta descripciones ampliamente detalladas en crónicas, diarios de viaje, obras históricas, reseñas geográficas. Asimismo los encontramos en noticias periodísticas, novelas, ensayos literarios, cuentos. Las observaciones y estudios sistemáticos se inician en el siglo XVIII y se deben a un variado conjunto de científicos, llámense naturalistas, historiadores, geógrafos, biólogos, arqueólogos, agrónomos, antropólogos sociales, etc. Estos materiales, aunados a los documentos primarios de archivo, constituyen un acervo considerable de gran riqueza que permite conocer tanto la evolución de la agricultura chinampera como el desarrollo histórico general de las zonas lacustres de la Cuenca de México, en un período de cuando menos quinientos años.

Entre los aspectos que mayormente destacan por la calidad y el detalle de la información, se hallan los de la tecnología de construcción de las parcelas, las obras hidráulicas, los cambios ambientales, los repertorios de cultivos y los métodos y técnicas agrícolas, incluyendo a los instrumentos.

En este ensayo presento únicamente el tema de la evolución de los repertorios de plantas cultivadas, tanto las nativas como las introducidas a partir del siglo XVI, desde la época prehispánica hasta nuestros días. Se trata de una síntesis de lo que consignan los más importantes de los registros antes mencionados, especialmente los de primera mano, así como de mis propios datos de campo, en relación con las especies, el modo de cultivarlas, los ciclos de producción, el destino de ésta, los cambios ocurridos en el período de estudio y los factores con los que éstos se relacionan. Al final de este ensayo aparece un cuadro en el que se virtieron datos sobre los cultivos contenidos en las diversas fuentes.

Las preguntas que se intentan responder, con ayuda de esta información, están relacionadas con la gran intensidad que se atribuye a la agricultura en chinampas, calificada como una de las más productivas del hemisferio occidental (véase por ejemplo Cox y Atkins 1979:127). Entre las interrogantes más importantes y que sirvieron de guía a esta investigación, están las siguientes: ¿Hubo un quehacer agrícola en el ciclo de heladas de otoño-invierno antes de la introducción de las plantas del Viejo Mundo? En el caso positivo, cómo se dio; cuántos ciclos agrícolas; qué plantas hubo al año; qué métodos lo permitieron; cómo se rotaron los diversos cultivos; cómo se asociaron. A partir de la adopción de las nuevas plantas traídas por los españoles, tanto de Europa como de Asia y Africa, cuáles fueron las formas, la secuencia, el ritmo y los efectos de la incorporación; cómo se combinaron con las nativas, a cuáles desplazaron, qué momentos de los ciclos ocuparon; cuáles fueron las causas de la selección (comerciales, ecológicas, demográficas, tecnológicas, culturales, etc.). ¿Desplazaron los nuevos cultivos a los mesoamericanos? ¿Se limitaron a "llenar" los vacíos en los que, quizá, no se cultivaba nada antes de la conquista? ¿Se sumaron acaso a las existentes, dando flexibilidad y variedad al repertorio total?

## Método

El método que se adoptó para sistematizar la información reunida e intentar responder a las preguntas y problemas anteriores, fue el de vaciar a fichas y a cuadros de síntesis los datos que se obtuvieron en las fuentes primarias y directamente en el campo respecto a cada una de las plantas cultivadas. Con frecuencia solamente se logró efectuar a nivel de conjunto, como en el caso de las flores, por ejemplo. La siguiente exposición sigue ese esquema por planta o por complejo de ellas, en un orden que refleja el de su importancia relativa. Se separan en dos grupos mayores, correspondientes a las plantas del Nuevo Mundo y a las del Viejo. El primero es el que se desarrolló con más profundidad debido a su mayor importancia en casi todo el período de estudio.

## Cultivos de las chinampas*

De las numerosas plantas nativas de Mesoamérica (o de antiguo cultivo prehispánico en el área) que se producen en estas parcelas, el maíz ha sido la principal y la más constante a través de los siglos. El jitomate le sigue en importancia, con una producción plenamente orientada al mercado; el tomate, los chiles y las calabazas también se cultivaron en forma predominante a fin de expenderse en la ciudad de México. Con excepción del jitomate, las otras hortalizas se han sem-

*Al final de este ensayo aparece una lista de especies con sus respectivos nombres científicos.

brado tanto en los terrenos de temporal de las partes bajas de los cerros aledaños a las chinampas, como en los de la llanura lacustre, en los cuales, antes de la desecación del siglo XX, se podían abrir pozos superficiales para la irrigación.

Frijoles, ejotes, chía, uauhtli (alegría, amaranto o bledo), chayote, chilacayote, diversas flores de ornato, hierbas de olor y "quelites" (hierbas comestibles de acuerdo con la tradición náuatl), completan la lista de los cultivos mesoamericanos en las chinampas, si bien algunos de ellos sólo han sido huéspedes temporales. Los chiles, la alegría, la flor de muerto o cempasúchil y la lechuga (introducida en el siglo XVI), pasan su primera edad en los almácigos colocados en la chinampa o en el "sitio" de la casa, pero maduran y fructifican en los suelos de los cerros o sus estribaciones en el período de lluvias, adonde son transplantados en cuanto comienza el temporal. Tanto magueyes como nopales eran cultivados abundantemente en los cerros y en la llanura lacustre de las zonas de Xochimilco y Chalco, hasta los primeros dos decenios del siglo XX aproximadamente, al igual que los olivos en la llanura, a partir de que fuesen introducidos en el siglo XVI (Gibson 1967:328).

Prácticamente todas las hortalizas así como gran parte de las flores y de las hierbas de olor traídas a América por los españoles durante la colonia, fueron incorporadas, paulatinamente, al repertorio de plantas cultivadas en las parcelas lacustres: pepino, coliflor, col, col de bruselas, alcachofa, ajo, cebolla, espinaca, acelga, cilantro, perejil, betabel, apio, nabo, rábano, poro, brócoli, zanahoria, salsifí, colinabo, etc. Cereales y leguminosas como trigo, cebada, alverjón (chícharo) y haba también han estado presentes alguna vez, pero su cultivo no parece haber prosperado debido, probablemente, al exceso de humedad en las parcelas. Entre las plantas de origen andino, traídas después de 1519 a Nueva España, únicamente se sembró la papa, pero, sobre todo como una "curiosidad".

De estos cultivos, conocidos a raíz de la conquista, se adoptaron algunos para producirlos en los cerros. Es una lista más limitada que incluye a los siguientes: haba, alverjón, papa, lechuga, lenteja, garbanzo y cebada.

Las flores, por las cuales destacan Xochimilco y su región, se han venido cultivando sin interrupción desde el siglo XVI. Hasta aproximadamente los años cuarentas o cincuentas del siglo XX se arreglaban para la venta en forma de "ramos" pero, a partir de ese período, se empezaron a preparar como "planta", es decir, en un cubo de tierra o en un bote para luego ser transplantados a jardines o macetas. Esta innovación se atribuye al cambio en las condiciones de humedad en las chinampas. Las más famosas fueron, sin duda, las amapolas, que lucían sus galas el "lunes de la amapola", que sigue al domingo de Pascua, cuyo cultivo fue prohibido por el Estado mexicano a principios de los treintas. Entre las flores que contaban con mayor preferencia, nativas e introducidas, se encuentran las siguientes: amapola, nube, chícharo, espuela de caballero, cempasúchil, alhelí, pincel, imperial, aretillo, nardo, pensamiento, alcatraz, dalia, violeta, crisantemo, etcétera. A continuación se mencionan las plantas más populares: mercadela, agazania, tuberosa, amapola japonesa, petunia, clavel, ester, chino, crisantemo, vara de San José, panalito, aretillo, rosa, etcétera. Por otra parte, en la actualidad se siguen cultivando las tradicionales, a saber: alhelí, nube, cempasúchil, pincel y chícharo.

Entre las principales hierbas de olor que se producen citaremos a la ruda, la hierbabuena, el romero y la albahaca. Se venden como planta en tanto que la manzanilla y la hierbabuena las presentan en forma de manojos.

Los árboles frutales, nativos e introducidos, dejaron escasa huella en los registros históricos; no obstante, se sabe que tuvieron alguna importancia en el siglo XX en algunos pueblos como San Luis Tlaxialtemalco, San Gregorio Atlapulco y Xochimilco. Las semillas se siembran en almácigo para comercializar el arbolito pequeño con su cubo de tierra y después se traslada al terreno definitivo donde se injerta. Esto ocurre en las tradicionales zonas productoras de fruta como son Huexotzingo y Juauhchinango, estado de Puebla. Las que más abundan son el tejocote, el durazno y el chabacano. En esta región se compra fruta corriente a fin de obtener la semilla y los arbolitos de Xochimilco sirven como "patrón" con el objeto de injertar variedades finas de frutales. Sin embargo, no se trata de una producción realmente importante. Algunos árboles adultos se quedaron en la orilla de las chinampas o bien en los patios de las casas, para consumo familiar de la fruta; entre ellos están el tejocote, el higo, el membrillo, el ciruelo y el durazno.

Maíz

La variedad que crece actualmente en las chinampas y terrenos húmedos de los antiguos vasos lacustres es el "chalqueño": tipo productivo, vigoroso y tardío al que los especialistas consideran de aparición reciente. Según ellos, esta raza de maíz comenzó a desarrollarse a partir de la época de la conquista, probablemente como resultado de la hibridación de las razas de cónico y tuxpeño (Wellhausen *et al*. 1951: 172). Sus límites superior e inferior de adaptación son 2300 y 2800 metros aproximadamente, y se localiza en terrenos irrigados o de humedad.

El chalqueño-chinampero es una variedad maicera de ciclo más largo que el de la que se utiliza en los cerros aledaños, que carecen de riego y de reservas subterráneas de humedad. Se siembra antes del inicio del temporal, en cuanto el peligro de heladas haya desaparecido (fines de marzo, principios de abril), gracias a la humedad contenida en los suelos; se cosecha hacia principios de noviembre. El chalqueño se encuentra ahora en todas aquellas tierras que tuvieron o aún tienen chinampas o bien en terrenos desecados de los antiguos vasos lacustres, como en Itztapalapa, Chalco, Tlalpan, Tláhuac, Mízquic, Xochimilco, San Gregorio, Santa Cruz, etcétera.

En las tierras de temporal acondicionadas con terrazas o metepantles (bancales) en las laderas de los cerros, se utiliza la semilla del maíz "cónico" o "maíz del cerro", como se le llama en la zona de Xochimilco. Sus límites de adaptación son más amplios que los del chalqueño, van de los 2800 a los 2200 metros. Es una variedad más precoz que se deposita en el suelo en cuanto se inicia el temporal (fines de mayo, principios de junio) y que se cosecha al comenzar el mes de noviembre.

La distribución geográfica del maíz cónico y del chalqueño en la Mesa Central de México es "casi idéntica", excepto por la altitud; pero, si bien ambos se encuentran en la misma región, no compiten por las mismas tierras sino que conviven uno al lado del otro (Idem:171).

En tiempos pasados, el cultivo del maíz era típicamente chinampero, con siembra en almácigo de lodo, trasplante, muchos cuidados. Es inexacto que la chinampa haya sido inadecuada para la producción del maíz y que sólo fuera propia para la de verduras, como algunos autores lo han afirmado (Olmedo, en base a Mejía Fernández; en Olmedo 1966, I:45). El padre Alonso Ponce, con su compañero Antonio de Ciudad Real, el autor de la crónica de su viaje, lo vieron crecer poderoso y exuberante en las chinampas de Xochimilco en 1585 (1976, I:108). Los mexica, en su peregrinaje en busca del asiento definitivo en la Cuenca de México, lo cultivaron en las chinampas de Xaltocan para, más tarde, una vez establecidos en Tenochtitlan, continuar haciéndolo en las numerosas chinampas que construyeron. Lo mismo sucedió con los habitantes de los señoríos vecinos durante la etapa de mayor expansión de la empresa de reclamación de estas parcelas en los pantanos de la Cuenca (Armillas la sitúa del siglo XIV al XVI; 1971:658).

En la zona urbana de la antigua ciudad, el maíz se cultivaba con el fin de obtener elotes, debido a la pequeñez de las parcelas (Zorita, en Scholes y Adams 1958:53; Calnek 1972:111). Sin embargo, en las zonas rurales de sus alrededores, lo mismo que en la subcuenca de Chalco-Xochimilco, se producía con el objeto de lograr maíz para grano, destinado al autoabasto o al tributo. Al respecto, Zorita escribió sobre Tenochtitlan:

> "Y si algunos siembran /maíz/, es muy poca cantidad en algunos camellones que tienen a la redonda de sus casas para lo comer en elote, que es cuando comienza a granar, que es como fruta, y lo tienen por mantenimiento." (Scholes y Adams 1958:53)

Las fuentes no indican el número de cosechas de elote que se obtenían, pero bien pudieron ser dos, con empleo de almácigos y abrigos.

Del uso de almácigo y el trasplante en relación con el maíz, hablaron tanto fray Bernardino de Sahagún como el mismo Antonio de Ciudad Real en el siglo XVI (Sahagún 1975:703; Ciudad Real 1976, I:108). A este último, que lo vio en Xochimilco, le pareció "... cosa muy particular de aquella tierra..." y lo describía de esta manera: "Ponen también en aquellas chinampas almácigos de maíz y de allí los trasponen..."

En el siglo XVIII, según la descripción hecha por Alzate, el almácigo para maíz lo acostumbraban los indios de Culhuacán y Xochimilco para producir cosechas "muy anticipadas a lo que se verifica respecto al maíz sembrado en el orden regular" (1791, en 1831, II:392). Lo mismo ocurría, como se expone más adelante, con las calabacitas, por ejemplo. El almácigo se cubría con techado, antes de trasponer las matitas, se les cortaban las hojas"...al modo que los agricultores lo ejecutan con las cebollas, ajos y otras plantas, y las siembras de maíz en las chinampas prosperan muchísimo: a la vista se presentan como unos bosques..." (1831, I:391-392). Es probable que estas siembras adelantadas de maíz se realizaran a principios de febrero, como aun ahora ocurre eventualmente, luego de la bendición de las semillas de maíz, frijol, calabaza, chile y chayote, el día 2 de la Candelaria

(2 de febrero).

Miguel Santamaria, agrónomo de tiempo de don Porfirio, describió detalladamente el cultivo del maíz al estilo tradicional, efectuado en el ciclo normal, es decir, como siembra adelantada. Comenzaba con la hechura del almácigo de lodo en marzo-abril, para terminar con la cosecha en noviembre. Antes del trasplante, al cabo de 37 a 45 días, se despuntaban o podaban las matitas dos o tres veces, a la manera descrita por Alzate. Para trasponerlas, éstas se desahijaban para solamente colocar en los hoyos o cepas las dos más vigorosas de cada cepellón ("chapín"); las líneas de hoyos se separaban entre si un metro y éstos, 84 centímetros. Enseguida se daba un riego "de asiento". En mayo se escardaban una o dos veces, entre junio y julio venía el aporque con azadón; luego de la fecundación, las espigas de las matas (inflorescencia masculina, <u>miauatl</u> en náuatl) se despuntaban para que las plantas se desarrollaran con más vigor. En noviembre ocurría la cosecha en la chinampa misma, dejando allí el rastrojo; el almacenamiento de la mazorca tenía lugar en colotes o sincolotes, "especie de huacales sostenidos por 4 o 6 pies derechos..." (Santamaría 1912:15-18).

Pedro Armillas recogió en 1946 el calendario de actividades seguido por un chinampero de Tláhuac, cuyas fechas se acercan a las anteriores, pero las operaciones presentan mayor sencillez. La siembra se hacía en marzo, en almácigo de lodo con chapines de 8 centímetros por lado, con 4 granos en cada uno; el trasplante en abril, en hoyos con estiércol en el fondo; en junio se efectuaba el montón o aporque, hecho con azadón; en agosto, el desyerbe con hoz, a partir de la segunda quincena de noviembre la pixca o cosecha (notas inéditas).

William Sanders recabó en 1953 un calendario para el cultivo del maíz tal como se hacía en los "viejos tiempos", quizá refiriéndose con ello a la época previa a la desecación. En febrero se preparaba el almácigo, en marzo se barbechaba el suelo con azadón y se fertilizaba con estiércol, agualodo y "huachinango" (lirio acuático, <u>Eichornia crassipes</u>) revueltos con la tierra. Al final de marzo venía el trasplante; en abril, mayo y junio se regaba dos veces a la semana en la etapa inicial (antes del comienzo de las lluvias); en abril se daba la primera labor o esda, en mayo la segunda escarda y el montón, en octubre la última labor; la cosecha se realizaba en noviembre (1957:82-83). En las chinampas recién construidas de San Luis, en las primeras décadas de este siglo se sembraba el maíz - puesto previamente en almácigo de lodo - en el "césped", el entretejido natural de plantas acuáticas llamado también "cinta", "pantano" o <u>atapalacatl</u>.

En San Luis observé en una chinampa que no tiene acceso por tierra, de tal manera que es imposible la entrada de la yunta, el barbecho con azadón, la hechura de hoyos a lo ancho de su superficie, cuyo trazo se hace con ayuda de un mecate con dos palos (se "raya con hilo"), así como el trasplante. El almácigo de lodo se preparó en marzo, con 3-4 semillas en cada chapín; se trasplantó en abril con todo y chapines, colocando en el fondo del hoyo abono de res en polvo y agualodo del fondo del canal.

En el resto de las chinampas y en los "terrenos" de las zonas desecadas de los antiguos vasos lacustres, se utiliza el azadón en combinación con el arado para barbechar, laborear y escardar; la siembra se hace directamente en marzo sin usar almácigo, con ayuda de la pala o la coa de hoja metálica. El uso del tractor para roturar y de los fertilizantes animales y químicos son cada vez más comunes en estas parcelas desecadas, ya sin canales ni árboles que estorban su paso.

La desecación ha simplificado no sólo las técnicas tradicionales chinamperas sino las formas de uso del territorio, de tal manera que puede afirmarse que casi todo el maíz que se cultiva hoy en las zonas de Chalco, Xochimilco o lo que aun es agrícola de la de Mexicaltzingo-Itztapalapa y Tláhuac, se trabajan con arado o tractor, o arado y tractor, sobre todo para barbechar o roturar el suelo. Con el arado, con el tractor o con ambos, se barbecha, cruza, rastrea, melga (surco), se laborea y escarda, con la pala o la coa se siembra, resiembra y se levantan las matas caídas en el laboreo. Con azadón se barbechan los "bordos" u orilla de los terrenos que fueron chinampas o bien aquéllas ya semidesecadas que conservan los ahuejotes. Con este instrumento muchos campesinos siguen laboreando, escardando y dando montón a las plantas del maíz. El desyerbe de agosto-septiembre lo realizan con hoz dentada, cortando la hierba sin remover el suelo.

Hace 25 años, cuando W. Sanders hizo su estudio en las zonas de Mízquic, Tláhuac y San Gregorio, las chinampas se trabajaban de la siguiente manera: al final de marzo se barbechaba con azadón , se sembraban 4 a 5 granos secos con la coa. Los surcos o líneas se espaciaban unos 80 centímetros y cada planta del surco, 50 centímetros aproximadamente. Se le daban una o dos labores y desyerbes para hacer el montón en junio. Esto último se hacía, al decir de los propios chinamperos, para que las plantas resistieran mejor las "tormentas" de la temporada de lluvias, características del final del verano. Las matas llegaban a medir a veces 4 a 5 metros de altura. La fecha de la cosecha era en noviembre, prolongándose a diciembre.

## La desecación y el uso de abonos

El ensalitramiento de los suelos es el efecto directo más relevante de la desecación general de los vasos lacustres. Dos son los efectos secundarios que destacan: en primer lugar, el empobrecimiento de los suelos y la consecuente necesidad de usar abonos, principalmente el estiércol de vaca, pero también el "químico"; en segundo lugar, la aparición y proliferación de plagas antes desconocidas o poco frecuentes. Las de mayor significación son el "chahuixtle" y las tuzas. El chahuixtle es descrito por algunos como un mosquito negro que seca las matas del maíz cuando la lluvia escasea. Sanders lo consigna como un hongo que hizo su aparición en 1953 en los resecos terrenos ejidales y en las chinampas, en ese año pobre en lluvias. Sus informantes tanto de Tláhuac como de San Gregorio le aseguraron que no conocían esta plaga y que nunca habían perdido una cosecha en los terrenos ejidales desde que comenzaron su explotación en 1924 (Sanders 1957:63-64).

## Rendimientos del maíz chinampero

No se tiene información del rendimiento que el maíz chinampero tuvo en los siglos pasados, excepto aquella que se puede inferir de una imposición tributaria hecha a los indios oficiales (artesanos) de la provincia de Xochimilco en 1558. Los que fueran vecinos "de la tierra", que habitaban en las tierras de temporal, tenían que trabajar en las milpas "donde se hiciere la sementera del rey", labrando un área de 5 por 10 brazas; y los vecinos "del agua", en las chinampas donde se hiciere la sementera, trabajar 1 por 10 brazas. De estas áreas o "medidas" saldría una fanega de maíz por cada tributario. La proporción era entonces de 1 a 5 de unas tierras respecto a las otras (Sobre el modo:106).

Otro dato se refiere al rendimiento del maíz en los campos de la hacienda de Coapa a finales del siglo XIX. Indica que era de 150 a 200 granos cosechados por cada uno sembrado (Figueroa Doménech 1899, I:360).

Santamaría hace sus cálculos sobre el rendimiento del maíz en las chinampas del Distrito Federal, no sin antes advertir al lector que los datos que los agricultores le dieron eran "muy inciertos y variados". Concluye que "el rendimiento medio es de 100 por 1, oscilando en 165 y 250", es decir, similar al maíz de la hacienda de Coapa (Santamaría 1912:19). La producción media por hectárea la calculó en 80 hectolitros, lo cual equivale a 8000 litros o kilogramos (8 toneladas).

Sanders se preocupó por este cálculo tan alto de Santamaría y decidió buscar sus propios resultados. Sólo uno de sus informantes de Tláhuac afirmó un rendimiento tan alto para las chinampas de los tiempos anteriores a la desecación (1957:84). Sus observaciones, mediciones e informes llevan a Sanders a considerar la cifra de 4000 kilogramos o 4 toneladas por hectárea, como la producción media en las chinampas de San Gregorio, cultivadas con maíz y, además, con verdura en forma intensiva. Su cálculo lo hace sobre la base de que un cuartillo de semilla pesa 1.5 kilogramos y que un bulto tiene 50 cuartillos de semilla. De acuerdo con informantes de Tláhuac, calcula esa misma producción para la época en que sus chinampas se explotaban intensivamente, antes de la desecación. Su conclusión general es la siguiente:

> "...el maíz en las chinampas produce un mínimo de 3.5 y más probablemente 4.0 toneladas por hectárea; con la salvedad de que bajo cuidados más intensivos, especialmente en cuanto al riego, este cálculo puede elevarse. Estos promedios representan los rendimientos más altos de maíz de ciclo largo del México moderno, con la excepción del maíz híbrido desarrollado experimentalmente por el gobierno, cuyas cosechas alcanzan hasta cinco toneladas por hectárea en suelos de altiplano" (Sanders 1957:84-85).

En las chinampas desecadas de Tláhuac obtuvo cifras similares de entre 6000 y 2500 kg./ha., máximo y mínimo en cuatro chinampas desecadas; o un promedio de entre 3700 y 4000 kg./ha. (Idem:84).

En el Mízquic actual, Venegas, empleando el método del Plan Puebla (1968), modificado por Campos (1972), encuentra que la producción de maíz, en las chinampas en un año, es de 1519.4 kg./ha. y 5098.4 kg./ha. La gran disparidad la explica por las diferencias en su manejo y cuidado, trabajadas con azadón y fertilizadas con abono químico. Muestreó siete chinampas en total, obteniendo la cifra promedio de 3.9 ton./ha. Cita cifras del CIMMYT (1974), con 3.13 ton./ha. con maíz mejorado, maquinaria, fertilizantes y pesticidas (Venegas 1978:8, 21 y 24).

Sanders examinó también los rendimientos del maíz en los terrenos ejidales de San Gregorio, Tláhuac y Mízquic. En una parcela "de primera" del ejido de San Gregorio, su informante obtenía de 15 a 17 cargas o 25 a 30 por hectárea (1800 (2250), 2040 (2550), 3000 (3950) y 3600 (4500) kg./ha., respectivamente, calculando 120 y 150 kg./carga), o sea, un promedio de 2610 (3312.5 kg./ha. (promediando las cuatro cifras). Al principio de la dotación ejidal y durante varios años, obtenían hasta 40 cargas/ha., es decir, 4800 (6000) kg./ha. Los terrenos eran "vírgenes".

En Tláhuac encontró una gran variación. Ofrece los datos de dos terrenos, uno de 2.4 hectáreas que rendía de 5 a 8 cargas/ha., o sea, de 600 (o 750) a 960 (o 1200) kg./ha., tomando 120 y 150 kg. por carga. El otro terreno rendía 16 cargas/ha., es decir, 1920 (o 2400) kg./ha. Cerca de Mízquic las parcelas rendían cantidades similares a las de las parcelas de primera de San Gregorio (Sanders 1957:64-65).

Las cifras oficiales para los ejidos de la delegación de Xochimilco indican un promedio de 1885 kg./ha. y 1973 kg. para el ejido de Tláhuac (citado en Sanders 1957:64-65).

Los esquilmos del maíz

Los frutos maduros del maíz, las mazorcas, han sido tan sólo uno de los productos que la planta ha rendido. Esta se muestra generosa con el hombre pues todas sus partes se pueden aprovechar, algunas incluso antes de que los frutos principales maduren. Los elotes, las espigas y las hojas han sido elementos económicos de gran importancia que han servido en más de una época de hambre. Así, las espigas se utilizaron hervidas para mezclarlas con los pocos granos de maíz que se tenían en los años de la Revolución de los 1910. Estas, las hojas verdes, la cañuela y el totomochtle secos resultaron excelente forraje para el ganado introducido por los españoles en el siglo XVI; los olotes, las "uñas" o raíces y la cañuela continúan siendo combustibles importantes en las cocinas de las casas campesinas. La cañuela, al igual que los tallos de la alegría, también se usan para "chinamilear" (cercar) una casa, una cocina o una chinampa. El totomochtle, elemento indispensable para envolver los tamales, se consume o se vende, proporcionando, en este último caso, pequeños ingresos a la unidad familiar.

Durante la época colonial hasta principios del siglo XX, las partes verdes o secas del maíz sirvieron para alimentar a los nuevos habitantes animales (caballos, mulas, bueyes, vacas) de México-Tenochtitlan y sus alrededores, los cuales fueron mantenidos con los esquilmos o restos del maíz y con la vegetación espontánea que crecía en las ciénagas y los cerros.

Desde principios de la dominación, los funcionarios reales de alta jerarquía y los habitantes españoles de la ciudad de México en general, tuvieron el privilegio de recibir de los indios de la Cuenca suministros diarios de "zacate" para sus animales; los primeros a la puerta de sus casas, a precios menores que los corrientes; los particulares en las acequias, a precios "moderados". Este suministro forzoso cargaba sobre los pueblos "de la laguna" o "nautecas" y sobre los "comarcanos". Los primeros llevaban zacate de la ciénaga o del maíz a los funcionarios, por "canoas" y a los vecinos se los vendían por "manojos". Los nautecas eran Iztacalco, Mexicaltzingo, Itztapalapa, Xochimilco, Culhuacán y Cuitláhuac (AGI, Escribanía de Cámara, 163 (A), f. 38v; CDHAI, IV:493-494). Con el tiempo, este comercio de forraje se convirtió en actividad primordial de estas poblaciones, en fuente de dinero y sustento ("Relación de Mexicaltzingo", 1580, en PNE, IV:194). Los pueblos comarcanos suministraban a los funcionarios coloniales, durante seis meses al año, cargas de hoja seca y verde de maíz. Estos pueblos eran en 1599: Coyoacán, Churubusco, Tacuba, Cuautitlán, Tenayuca, Tultitlán, Tepozotlán, Coyotepec, Teoloyuca, Huehuetoca, Zumpango, San Cristóbal (Ecatepec), Chiconautla, Chimalhuacanejo y Tacubaya (Gómez de Cervantes 1944:113-114).

En las zonas lacustres las hojas del maíz se recolectaban entre septiembre y octubre. En 1791 Alzate dio cuenta de su tráfico por el canal de Mexicaltzingo (de la Viga) en estos meses: "Hojas de la planta del mais que se transporta para el sustento de las bestias." (1831, II:306).

La actividad de recolectar y vender pastura, iniciada como un servicio forzoso en el siglo XVI, continuó hasta principios del XX. Esquilmos y "zacate" de la ciénaga sirvieron de alimento a las vacas lecheras de los establos situados en distintos puntos a lo largo del canal de Mexicaltzingo, como Santa Anita, Iztacalco, etcétera. Mucha gente de los antiguos pueblos "nautecas" continuó dedicada a recolectarlos y venderlos a los "contratistas de zacate" locales que los revendían a los estableros. Según Santamaría, los "zacateros" de su época (1912), a quienes por cierto distingue de los trabajadores cultivadores que eran propietarios, constituían un grupo de individuos "...que á las dos o tres de la mañana se van á las chinampas ajenas á cortar las hierbas, hojas de col y todo lo demás que pueden, haciendo manojos que venden á los contratistas de forrajes; de este modo se hacen de 75 centavos ó un peso diariamente, sin trabajar más de tres horas; el resto del día lo pasan en las tabernas" (1912:36-37). Efectivamente, los zacateros o dedicados "al ciento" (de manojos), eran la gente más humilde de los pueblos, los que también se dedicaban a "la vara", a conseguir varas de árboles de ahuejotes o de otros para la elaboración de canastas y chiquihuites, que también se hacían de carrizo, o a hacer petates de tule ancho. Estas fueron especialidades tanto de Tulyehualco como de San Luis hasta principios del siglo XX, al parecer. Eran épocas en las cuales en varios de los pueblos lacustres había muy pocas chinampas en explotación.

## El maíz y otros cultivos

En relación a las asociaciones de otros cultivos con el maíz, se observa que se han hecho tanto en el momento mismo de la siembra (concomitante), como en momentos distintos (imbricada), o bien en mosaico (en camellones). En los primeros años del presente siglo se asociaba maíz, frijol y chile en una siembra concomitante, hecha en el momento de trasplantar el chile del almácigo al suelo de la chinampa, en el mes de marzo "...entre cada dos líneas de chile se intercala una de maíz y el frijol se siembra en los entresurcos. Se cosecha el chile y queda ocupada la chinampa por frijol y maíz, cuya recolección tiene lugar más tarde" (Santamaría 1912:20-21). La cita parece indicar que tanto el maíz como el frijol se sembraban directamente, sin usar almácigo.

Los chinamperos de San Luis y Mízquic recuerdan una asociación del frijol o del uauhzontle con el maíz en la misma chinampa. El frijol se sembraba al pie de los ahuejotes para que la guía se enredara en ellos, el uauhzontle en las orillas de la parcela. Ambos cultivos ocupaban el suelo al mismo tiempo que el maíz aunque por menos tiempo. En San Luis se sembraba col en medio del maíz y el chile (ciclo marzo-noviembre), en el mes de julio, la cual quedaba en el terreno una vez cosechado el maíz (ciclo total de la col: seis meses). Hoy se asocia maíz con apio, o maíz con espinaca, si el terreno "esta bueno". El apio se ensemilla primero en semillero de tierra, "regando" la semilla; se pasa luego a un almácigo de lodo, para trasponerlo al suelo de la chinampa entre los surcos del maíz. Debido a que su ciclo es más largo (un año), permanece en el terreno después de cosechado el maíz. En el caso de la espinaca, su semilla se pone en almácigo de lodo en marzo, para luego trasponerla entre los surcos del maíz e iniciar los cortes en junio.

Respecto a la rotación del maíz con otras plantas, se tiene el registro de Sanders referido a San Gregorio en los años cincuentas que, según mis datos, no parece haber ocurrido en San Luis, quizá por la distinta forma de orientar la producción en los dos pueblos: más comercial en el primero a causa de las mejores condiciones del área de explotación y a la existencia del ejido para el cultivo del maíz. Dice que la situación más común respecto a la producción de este cereal en las chinampas de San Gregorio, era la de sembrarlo únicamente un año enseguida de dos a cuatro años de cultivar sólo verdura. La última cosecha de ésta ocurría en diciembre, el ciclo de maíz iba de marzo a noviembre del año siguiente. Se decía que con el maíz la tierra "descansaba" de la intensa producción de hortalizas (Sanders 1957:80-82). También se rotaba maíz con jitomate y chile, un año el primero, al otro los segundos (Idem:89).

En San Luis, en tiempos pasados, hacia los inicios del siglo XX, se rotaba maíz con lechuga orejona: en octubre se recogía el maíz en mogotes para dejar libre el terreno para trasplantar la lechuga ese mes (el almácigo se ponía en septiembre). Se cosechaba en diciembre, enero y febrero. La principal rotación que hoy se practica con el maíz en las chinampas en buen estado o en las semidesecadas con humedad subterránea, es con el romerito que se expende en el ciclo de Cuaresma. El romerito se siembra directamente en el suelo en el mes de octubre, regando la semillita, pero debido a que el maíz aun no arrecia lo suficiente para poder cosecharlo, se cortan sus matas, se amogotan para dejarlo secar allí; mientras se cultiva el romerito en el espacio desocupado, en camellones. Sólo a veces se rotan romeritos con calabacitas.

Jitomate, tomate, chiles y calabazas

Su importancia decayó apenas en el siglo actual, con la desecación y los cambios en las redes de abastecimiento de la cada vez más grande y poblada ciudad de México.* El jitomate ya no se cultiva actualmente en las chinampas en tanto que el tomate, los chiles y las calabacitas nada más en pequeña escala en unas cuantas parcelas. Los cuatro mencionan sistemáticamente en los registros históricos de todos los siglos a partir del XVI (véase cuadro-resumen). Sin embargo, la primera descripción detallada de su proceso de producción la hizo hasta 1791 el sabio de Ozumba, José Antonio Alzate.

El conjunto de los datos indica que fueron los productos de mayor improtancia comercial en la economía de las zonas chinamperas, cuando menos desde fines del siglo XVIII en que esta documentado, hasta antes quizá de que fueran sustituídos casi por completo por las hortalizas del Viejo Mundo, entre 1930-1950 (fechas probables), las cuales siem-

---

*Las siguientes cifras de población de la ciudad de México ilustran el ritmo de su crecimiento desde el siglo XVII: 1600:58 500; 1700: 105 000; 1800:137 000; 1845:240 000; 1900:541 000; 1910:721 000; 1921:906 000; 1930:1 230 000; 1940:1 760 000; 1953:3 480 000; 1960: 4 374 000; 1970:6 874 165.

pre estuvieron presentes desde que fueron introducidas. Los chinamperos las usaron para combinarlas con las plantas nativas, pero acabaron por desplazar a éstas casi por completo, en especial al jitomate y al tomate.

En los siglos XVIII y XIX la ciudad de México era abastecida de verduras y flores que provenían de tres zonas productoras: 1) de las chinampas de Iztacalco, Xochimilco, Tláhuac y Chalco; 2) de las huertas de San Cosme, establecidas poco después de la conquista a lo largo de la calzada México-Tacuba; 3) de la "tierra" o "tierras calientes", referencias probables a los valles de Cuernavaca, Cuautla y Chietla. Las tres aprovisionaban, coincidiendo algunas veces con el abasto de determinados productos vegetales, turnándose otras. Fue así que las calabazas tiernas, el chile fresco, el tomate y el jitomate de las tierras calientes entraban al mercado de México de noviembre a marzo, precisamente durante los cinco meses en que hiela con más fuerza en la Cuenca de México. Entretanto, los productores del Valle introducían "vituallas de todas especies", coles y lechugas (Alzate 1831, II:298). Los de las chinampas se expendían de abril a junio (Santamaría 1912: 19-20), antes de que las tierras de temporal de la propia Cuenca produjeran sus frutos, a fin de alcanzar buenos precios al no tener contrincantes, si bien terminaban el ciclo compitiendo con los de las tierras de temporal (Alzate 1831, II:297-298; AGN, Desagüe, vol. 43, exp.5).

No obstante que las cuatro plantas de este apartado tienen en común su poca resistencia a las heladas, su cultivo tuvo lugar en diversas épocas del año, incluyendo la de invierno. Con el jitomate se practicó lo que Santamaría llamó con exactitud siembra "forzada", en tanto que con las calabacitas y el chilacayote, lo que los campesinos de hoy califican de siembra "adelantada" o "anticipada", puesto que su ciclo se inicia un poco a contrapelo, cuando todavía no cesan las heladas.

<u>El jitomate</u> Este se producía en la temporada de secas, pese a ser la más fría del año y con el riesgo de que sus frutos se "quemaran" con el hielo. Ello se debía a que el jitomate se enfermaba (pudría) al contacto con el húmedo suelo de la chinampa, de tal suerte que su ciclo tenía que desfazarse para que no coincidiera con el período de lluvias. De esta manera el jitomate de las chinampas tampoco competía con el producido en las tierras calientes temporaleras. Lograr esta siembra forzada conllevaba complejos cuidados así como proporcionar atención individual a las matas: empleo de uno y hasta dos almácigos, abrigos con tolchimales o tapaderas, bordos de tierra para protegerlas del hielo; la práctica de aflojar, el riego y las escardas. Aquellos que lo observaron se sorprendieron de la gran inversión de mano de obra y de la delicadeza que los chinamperos eran capaces de desplegar para producirlo. Santamaría escribió, muy significativamente, lo siguiente: "...es el que recibe mayores cuidados de los cultivadores, si bien es cierto que les reporta muy buenas utilidades" (1912:19-20). Y Sanders, (su cultivo) "...da una buena idea de la intensidad del cuidado dado a las cosechas y de la cantidad de trabajo invertido en su producción" (1957:92).

Santamaría describió este ciclo con tanto esmero como el que la planta recibía de los campesinos de Xochimilco en 1911. Se puede resumir de la siguiente manera: los tres primeros meses permanecían las plantitas, primero en el semillero de lodo (un mes a partir de septiembre), luego en el almácigo (dos meses a partir de octubre), donde se le prodigabanenormes cuidados, sobre todo por medio de los abrigos que las protegían tanto del sol y las lluvias que aun caían en septiembre, como de las heladas nocturnas a partir de octubre. Cobertizos de ramas servían para proteger al semillero y "tapas" o "tolchimales" de tule o cañuela, o bien de grueso césped seco (pantano o atapalacatl), sostenidos con carrizos, para cada cuatro plantas del almácigo.

Para contener el crecimiento desmesurado de las matas antes del trasplante, así como para evitar la disminución posterior de la fructificación, se desenraizaban "de vez en cuando". A esto le llamaban "aflojar" (según Santamaría), o "acomanear" o "poner en acomanas" (según los actuales chinamperos). (Del náuatl ahcomana: alborotar, Luis Reyes, comunicación personal.)

A partir del trasplante hecho en diciembre, además de un riego "de asiento" inicial, se volvían a proteger las plantas del jitomate con los tolchimales y los carrizos, escardando una o dos veces (ver los dibujos en Santamaría 1912). La cosecha se realizaba en abril a junio. También durante ella continuaban los cuidados; se extendía en el suelo "...una capa espesa de tule bien seco para evitar que se pudran los jitomates por su contacto con el suelo húmedo" Se recogían "...cinco cortes abundantes, uno cada quince días y dos o tres secundarios para recoger los últimos frutos" (Santamaría 1912:20).

En 1946 Armillas registró el calendario seguido por un chinampero de Mízquic para el cultivo del jitomate. En noviembre sembraba el almácigo, en chapín del mismo tamaño que para el maíz, es decir, de unos 8 centímetros por lado; protegía a las plantas de los hielos con "petatera", un abrigo hecho de tules amarrados con mecate que podía enrollarse. En enero trasplantaba y continuaba abrigándose con "tapadera de césped" o de "zacate de monte", así como con los riegos manuales con agualodo aplicado con zoquimaitl o "cuero". Estos se hacían necesarios hacia mayo, cuando las plantas estaban más grandes. La cosecha ocurría durante junio y julio (notas inéditas de P. Armillas). Según este mismo observador, el tomate se sembraba en Mízquic en pequeña escala en la misma chinampa que el jitomate, pero sólo en el borde. No registra la sucesión jitomate-tomate que Santamaría documenta para Xochimilco en los principios del siglo XX.

La gente en Xochimilco nos indicó que el jitomate se sembró allí hasta 1940 aproximadamente. La desecación que ya entonces predominaba acabó con su cultivo. Se tiene el siguiente calendario obtenido de los chinamperos que nos describieron su cultivo: el almácigo se ponía en noviembre con uso de tolchimales, se trasplantaba en enero a "surcos" (¿los "camellones de San Luis actual o los "surcos de pantano" de antes?), que medían un metro de ancho y de largo el ancho de la chinampa; cada uno con su bordo de tierra; su suelo se preparaba con "basura" (vegetación acuática) y una capa de lodo. Cada "surco" se separaba del siguiente con una distancia de tres metros para que la guía del jitomate tuviera espacio para extenderse. En cada "surco", lo más hondo posible, se colocaban las plantitas, separadas entre sí unos 30 centímetros. Por las tardes se colocaban las tapaderas, una cada seis matas, para protegerlas de las heladas; si éstas arreciaban, el bordo se levantaba, a veces hasta 20 centímetros. Así "sufrían", según la expresión de los viejos chinamperos, durante dos y medio o tres meses.

El fruto aparecía en marzo, se cosechaba en mayo y junio.

En las chinampas nuevas de San Luis, que se hicieron de 1914 a 1940 aproximadamente, los chapines de jitomate se colocaban encima del césped; se cubrían con cucuruchos de tule o zacate y no con tapaderas; igual ocurría en Mízquic. Se instalaba el almácigo en enero para cosechar en tiempo de aguas (verano).

Especialmente en Tláhuac y San Gregorio continuaba el cultivo del jitomate todavía hacia 1953. Su producción en estos sitios fue descrita con detalle por Sanders, de acuerdo con lo que le indicó un informante de Tláhuac. El, al igual que el que informó a Armillas en Mízquic y a nosotros en Xochimilco, Mízquic y San Luis, no utilizaban sino un único almácigo de lodo con chapines. Las fechas del ciclo también difieren respecto a las de Santamaría y Alzate pues no se sembraba en septiembre o noviembre, sino en diciembre. No se ponía en la chinampa sino en el patio de la casa para evitar robos. El suelo al cual se trasplantaban las matas en marzo se barbechaba con azadón y se fertilizaba con vegetación acuática. Después se labraban pozos en cuyo interior se echaban plantas acuáticas y agualodo para recibir a las plantitas. Luego venían los riegos con agualodo cada tres o cuatro días y sólo una vez a la semana cuando ya estaban bien desarrolladas. Simultáneamente se escardaban y laboreaban varias veces las matas y en mayo se podaban las hojas inferiores a fin de que pudiera aumentar el follaje.

En junio comenzaba la fructificación, pero los cortes o "chalipas" importantes ocurrían en julio y agosto, una cada ocho días. Durante este período, el de lluvias más intensas (junio-agosto), las plantas eran cubiertas con vegetación seca con el objeto de protegerlas del sol y la lluvia. También en este momento tenía el dueño que vigilarlas constantemente para prevenir los robos, para lo cual se instalaba en la chinampa, armado con su escopeta. Cada corte semanal rendía una canoa de uno por seis metros (Sanders 1957:92-93).

En esos años cincuentas, como ya se apuntó, el jitomate y el chile se rotaban con el maíz en la misma chinampa; un año maíz, al otro jitomate y chile (Sanders 1957:89). En la actualidad ya no se cultiva el jitomate de manera que nuestra búsqueda de semillas "criollas" del mismo fueron estériles. Lo describen como un fruto grande, achatado en los polos.

Las calabazas y el chilacayote Su cultivo en las chinampas y en los cerros circundantes es de antigua tradición. No se sabe con exactitud qué especies o variedades de calabazas hayan sido producidas, pero probablemente fueron algunas de las que los autores actuales han consignado y que aún hoy se cultivan: la calabaza "de Castilla"; la calabacita redonda, "criolla", "de codito", "estrellita" o "italiana" y la alargada o "japonesa". Estas dos últimas se venden frescas (razas de *Cucurbita pepo* L.; la primera de C. *moschata* Duech. o C. *mixta* Pang.)

Los chinamperos de los siglos XVIII, XIX y XX, cuando menos, ensemillaban a principios de febrero, para vender calabazas hacia mayo y junio, antes de que entraran al mercado de la ciudad de México las de las zonas de temporal. Las cultivaban temprano para venderlas caras, como me dijo un chinampero de San Luis. Lo lograban con los mismos métodos que el jitomate, es decir, con el uso de almácigos y abrigos que protegían a las plantitas de los hielos. Alzate los describió así: "...la semilla de la calabaza la siembran en almácigo de cieno...; en poco terreno se siembra mucha semilla y se cubre con facilidad para libertarlo de las heladas; de aquí depende el que los indios que practican este ramo de agricultura, vendan por mayo y junio pequeñas calabazas..." (1831, II:391). Los cuadrados donde las semillas se colocaban eran de tres dedos (unos 5 cm.) (Idem:393).

Un chinampero de Xochimilco me indicó que el calendario que seguían para el chilacayote era el de instalar el almácigo en noviembre y sembrar en enero y febrero, teniendo que usar tolchimales. El cultivo adelantado de calabaza y chilacayote se realiza aún en forma muy limitada. Lo vi en Mízquic, donde continúa el uso de tapar las matas con tolchimales de cañuela, pero ellos mismos dijeron que antes se usaban unos cucuruchos hechos de "zacate tres esquinas". Se ponía el almácigo a fines de febrero para sembrar a fines de mayo. En San Luis se cultivan actualmente calabazas usando almácigo que se prepara en febrero o marzo, para trasplantar en marzo-abril. No se realiza ninguna rotación especial de cultivos, si bien en una chinampa observamos, por ejemplo, calabaza durante la temporada primavera-verano y romerito en el invierno.

Respecto a su importancia relativa en los principios del siglo XX, Santamaría las sitúa entre los cultivos de menor relevancia (1912:26), en tanto que Sanders las ubica entre los diez más significativos de San Gregorio en 1953 (1957:77-78).

El tomate Al igual que el jitomate, el tomate fue un importante cultivo de las chinampas, del cual todos los datos parecen indicar que se produjo en más de un ciclo al año. Si se toman en cuenta los calendarios consignados por Santamaría a principios del presente siglo, resulta que el tomate verde, a diferencia de la práctica seguida con el jitomate, se prefirió sembrar en época de lluvias a fin de suceder a aquél en la misma tierra. No obstante que los ciclos de ambos suman trece meses, el empleo de almácigo permitió la rotación y por lo tanto, el uso continuo de un mismo pedazo de suelo. En San Luis nos dijeron que el tomate se producía en cualquier tiempo en el que no hubiera hielo.

Según Santamaría, se ensemillaba en almácigo de lodo a principios de agosto, el mes de lluvias más intensas del año, protegiéndose con tule seco. "...la cantidad de semilla empleada es de 2500 kilogramos por hectárea". A los veinte días, cuando las plantitas medían de 6 a 7 centímetros, se hacía el trasplante al suelo definitivo, en líneas separadas entre sí por una distancia de 84 centímetros para cosechar los frutos en los primeros días de noviembre (Santamaría 1912:21-22). Es decir, para cubrir la demanda de los días de "los muertitos", de Todos Santos (fines de octubre-principios de noviembre), aprovechando que ya en esos meses alcanzaba buenos precios. A los pocos días del trasplante se intercalaba nabo en los entresurcos del tomate, sembrándolo "al voleo y tupido"; éste se cosechaba dos meses después, en octubre (Santamaría 1912:21-22).

Ahora bien, si se toma en cuenta el calendario de Alzate, resulta que el tomate se cultivaba, al igual que el jitomate, en tiempo de heladas. En octubre se hacían los almácigos de lodo con chapines de dos dedos y medio (4 cm. aproximadamente). No consignó la fecha de la cosecha, pero es probable que fuera unos tres meses después, a principios de enero (de acuerdo con el ciclo señalado por Santamaría).

Un documento del siglo XIX nos da cuenta de su presencia en otra temporada: el mes de mayo, en las chinampas de Iztacalco y sus alrededores (1826; en AGN, Ayuntamientos, col. 10, f. 237r-274v).

Todo lo anterior confirma que la producción del tomate, a diferencia de la del jitomate, no estuvo sujeta a un período tan preciso del año, limitado por condiciones ambientales, sino que pudo ajustarse a la demanda o a otros factores sociales. Esta flexibilidad permitió su rotación con otros cultivos así como el uso del suelo en forma contínua (en rotación con el jitomate o con otros), aunque procurando que su cultivo no coincidiera con las épocas de heladas ni con las de la competencia con otras zonas productoras (especialmente las de temporal que entraban de junio a julio).

Al igual que el jitomate, el tomate se produjo en gran escala en la zona de Xochimilco, y en menor escala en la de Iztacalco-Mexicaltzingo, según datos de principios del siglo XX. No vi tomate cultivado en las chinampas durante mis recorridos, tampoco Sanders lo observó en 1953; no obstante, West y Armillas sí lo reportan en la década anterior. Actualmente se siembra en pequeña escala en los cerros de la zona, entre Xochimilco y Mízquic, sin pasar por las chinampas, salvo en algunas ocasiones en las que se opta por ensemillarlo en almácigo para luego trasplantarlo al cerro (Venegas 1978:19). Ahora, por lo general, es un cultivo de temporal cuya semilla se coloca directamente en los terrenos barbechados y surcados con el arado en forma similar a la que se sigue para el maíz o la alegría. Casi siempre se produce para sacarlo a la venta en Todos Santos.

Probablemente como continuación de una práctica prehispánica, el tomate y el chile de chinampas y cerro se fertilizaban de preferencia con guano de murciélago más que con vegetación acuática como las demás especies en las chinampas. A ambas plantas, se dice, les gusta "lo caliente", por ello en el cerro se dan muy bien puesto que allí las tierras son arenosas y "calientes" y el guano de ratón viejo también lo es.

El chile Era cultivado constantemente en las chinampas, pues si bien es víctima de las heladas como los otros, su producción fue posible gracias a los cuidados que los chinamperos le prodigaban durante el invierno. No sufre con las lluvias pero sí con el granizo que daña los frutos. Hoy se siembra en pequeña escala en las chinampas y en los cerros y en el termporal es trasplantado desde semilleros de lodo instalados en las chinampas o en los "sitios" (patios) de las casas.

Las fuentes sólo distinguen algunas de sus variedades conformándose con apuntar que se sembraban chiles o "pimientos", a los que Humboldt se apresuró a identificar como *Capsicum*. A continuación aparece una relación de los nombres comunes documentados: *chílchotl* o chile verde; criollo o chicostle; tornachile; chilito; carricillo; poblano; pasilla; largo; menudo; mulato y cascabel. Es decir, variedades tanto de *Capsicum annum* L. como de *C. frutescens* L.

En su peregrinaje a Tenochtitlan, los mexica sembraron en las chinampas de Zaltocan el chilchotl, que Molina tradujo como "ají o chile verde", que probablemente sea el que hasta hoy llamamos serrano o verde. Alzate lo consignó como "chile verde" o "pimiento pequeño" (C. annum). Otra variedad es el "tornachile", el tonalchilli que Molina traduce como chile "de regadío, que se hace en tiempo que no llueve" (1970); los chinamperos actuales lo describen como de color verde limón. Similar en forma al poblano o al güero. Tornachile y chilito estaban en las chinampas de Iztacalco y sus pueblos en 1826, en el mes de mayo en el que cayó una granizada inesperada (AGN, Ayuntamientos, vol. 10, f. 237r-274v). El carricillo, el criollo o chicostle, el poblano y el pasilla se sembraban en Itztapalapa y San Luis de acuerdo con la información de los campesinos que entrevistamos.

Las principales variedades producidas en Xochimilco e Itztapalapa a principios de este siglo eran el tornachile, el poblano, el largo y el menudo; en menor escala, el cascabel (Santamaría 1912:20-21). Según Alzate, en 1791 los indios de Xochimilco sembraban mucho tornachile; Santamaría para 1911 lo apunta entre los cultivos predominantes tanto de esta zona como de la Itztapalapa (1912:14-15).

El ciclo de cultivo para el chile registrado por Alzate es el siguiente: el almácigo se hacía en los días finales de septiembre, se ensemillaba en cuadrados de dos dedos de ancho (3 cm. aproximadamente), luego se le colocaba una cubierta que lo protegía de las heladas (Alzate 1831, II:393). West y Armillas dicen que las semillas se colocaban en chapines de 4 a 5 centímetros y las plantitas se trasplantaban a los noventa o cien días (1950:177). Para el chile y para el tomate se utilizaba el estiércol de "ratón viejo" o "chinaca". Si bien el registro más antiguo al respecto es el de Alzate, es probable que el uso de este abono sea mucho más antiguo, quizá de la etapa prehispánica. En el siglo XVIII se obtenía de las "dilatadas cuevas" del cerro de Itztapalapa, en las cuales se albergaba "un número infinito de murciélagos" que al anochecer salían formando "unas columnas interminables, que la paciencia en observarlas se fatiga", como gráficamente lo describió Alzate (1831, II:306). Posteriormente, los chinamperos lo consiguieron a través de personas de la zona de Xochimilco y Tulyehualco que dedicaban parte de su tiempo (de enero a mayo) a recorrer cuevas y conventos coloniales abandonados con el objeto de recolectarlo, en los estados de Morelos, Guerrero, Puebla y Oaxaca. Se habla de algunos que se dedicaban a este oficio en forma permanente.

En el siglo XVIII los cultivadores de chile de Xochimilco arrendaban las cuevas del Cerro de la Estrella con el fin de extraer el fiemo que luego mezclaban con tierra, probablemente con la del terreno al cual se trasplantaban las plántulas del almácigo. El uso de este guano para fertilizar chile y tomate (en las chinampas) y maíz, chile y tomate (en los cerros y planes) continuó hasta hace pocos años, cuando se extendió el de gallina así como los fertilizantes químicos. Aun hoy, sin embargo, la gente reconoce que el mejor abono para el chile es el de murciélago.

Santamaría consignó dos ciclos para el picante. El primero de diciembre a abril, que seguían aquellos campesinos que poseían tapas o abrigos para cubrir las matas que hacían posible su cultivo "forzado" en tiempo de heladas y la consecuente cosecha "adelantada". El segundo, de marzo a julio, es decir, un poco antes y durante el temporal, cuando las heladas "ya son poco frecuentes" (1912:20-21). El trasplante se realizaba en surcos separados por un metro, con una distancia de 84 cm. entre cada planta. Además de los riegos necesarios, se le daban dos o tres escardas para finalemnte obtener "...por lo regular de cuatro a cinco cortes abundantes, uno cada 9 ó 10 días" (Santamaría 1912:20-21). Ello ocurría de mayo a julio. Tornachile y chilito, como ya se mencionó anteriormente, estaban sembrados en las chinampas de Iztacalco el día de la granizada, el 24 de mayo de 1826 (AGN, Ayuntamientos, vol. 10, f. 237r-274v).

En Xochimilco nos indicaron que el tornachile no era de almácigo sino que se sembraba directamente, en "surco de a cuatro" (matas), de un metro de ancho, similares a las descritas para el jitomate. Ocurría en febrero, usando tapaderas; lo sembraban en marzo-abril, lo cosechaban a partir de julio. El terreno lo preparaban con lirio acuático y encima una capa de lodo en la cual hacían los hoyos para colocar los chapines con ayuda de un "picador" (instrumento de madera con la punta en forma de trompo). Se irrigaba con agualodo durante la temporada previa al comienzo de las lluvias.

El chile se asociaba con el maíz y el frijol, como ya se mencionó. Estos últimos, apuntó Santamaría "...se siembran al mismo tiempo que se trasplanta el chile /marzo/; entre cada dos líneas de chile se intercala una de maíz y el frijol se siembra en los entresurcos. Se cosecha el chile y queda ocupada la chinampa por frijol y maíz, cuya recolección tiene lugar más tarde" (1912:20-21). En la actualidad se continúa sembrando chile en doble hilera o "surco doble", pero ya no se asocia.

La costumbre de hoy de "echar a nacer" en los almácigos de la chinampa las semillas de chile u otras como alegría, cempasúchil o lechuga, por ejemplo, para trasplantar las matas a los terrenos de temporal de los cerros, no está documentada para siglos anteriores al XX; sin embargo, es posible que sea mucho más antigua. El chile en especial, pero también el tomate, se cultivaron en terrenos de las estribaciones de los cerros así como en los "planes" (planicies) lacustres, en los cuales se perforaban pozos de poca profundidad para irrigar a brazo. Los campesinos más viejos a los que se les consultó sobre este aspecto, informaron que, con la sola excepción de la alegría, las otras plantas se pueden trasplantar indistintamente del almácigo a la chinampa o bien al cerro. A finales del siglo XIX y durante las primeras décadas del XX, en San Luis se instalaban los almácigos de chile en los terrenos conocidos en esta región como "corrales" y que se hallaban situados entre las zona de chinampas y el plan, en los cuales existían pozos para regar. Trasponían en marzo y cosechaban desde junio hasta noviembre.

## Frijol, chia, uauhtli, flores y quelites

Estas plantas completan la lista de las de origen mesoamericano que se cultivaron en las zonas chinamperas. No dejaron más huella en los documentos que breves menciones (véase el cuadro-resumen). Unicamente del cempasúchil, al que Alzate se refirió de paso, encontré hasta ahora una descripción detallada del proceso de producción.

La chía (*Salvia hispanica* L. o *Hyptis suaveolens* Pot o quizá ambas) aparece citada en algunos documentos coloniales tempranos pero no se conoce hoy en la zona, ni tampoco la recuerdan los chinamperos más viejos. No obstante, existen datos del siglo XVI, referentes a otras zonas de la propia Cuenca de México, que indican que su cultivo era frecuente en el área.

El frijol chinampero (Phaseolus vulgaris L.) era una planta trepadora, de semilla grande. Actualmente se siembra en forma muy esporádica en las chinampas, pero todavía hace una o dos generaciones, era común observarlo durante el temporal, creciendo al pie de los ahuejotes que bordean las parcelas, para que en ellos enredara su guía. El ejote de frijol canario o "guachinango" se produce poco; se siembra en el mes de marzo. Otras variedades de frijoles de mata y enredadores, tanto "delgados" (P. vulgaris) como el "gordo" o ayecote (P. coccineus L.), se producen hoy en día en los cerros, pero sus semillas no se intercambian del cerro a la chinampa. En el cerro el terreno se prepara en forma similar a la acostombrada para el maíz, con yunta y arado, recibiendo beneficios similares. El frijol chinampero se sembraba directamente en el suelo en marzo o abril, con coa, poniendo un poco de lodo del canal en el hoyo. Se cosechaba al mismo tiempo que el maíz, entre noviembre y diciembre. En San Luis lo consideran muy productivo, que "da mucho", tanto que en ocasiones tenían que colocar puntales para sostenerlo. Se dice que se perdió debido a la falta de agua así como a las plagas que infestan los ahuejotes (Malacosoma azteca o gusano de bolsa, entre otras).

De los quelites (del náuatl quílitl), verduras o hierbas comestibles (especies de los géneros Amaranthus, Chenopodium, Porophyllum y Rumex, principalmente), se continúa sembrando el uauhzontle o quelite mazorca (Chenopodium nuttaliae Saff.), el epazote (Chenopodium ambrosioides L.) y el romerito (en náuatl iztacquilitl o quelite salado) (Suaeda torreyana). Los dos primeros son de temporal, en tanto que el tercero es una de las pocas plantas comestibles de origen mesoamericano que es cultivada en las chinampas durante la época de heladas. El uauhzontle y el epazote se siembran en almácigos o bien directamente en el suelo, "regando" la semilla; cuando las plantitas estan listas para trasplantarse, se "aflojan" para pasarlas posteriormente al terreno definitivo. De estos quelites se dice que son "perros", es decir, que son poco exigentes, que "agarran" donde sea. El romerito se produce en dos temporadas ligadas al calendario religioso y a la vigilia: la de Cuaresma-Semana Santa y la de Nochebuena-Navidad. Su cultivo se describió en el apartado sobre el maíz y sus asociaciones.

La planta de la "alegría" (Amaranthus leucocarpus S. Wats.) de la cual se obtienen las diminutas semillas de uauhtli para hacer el dulce de alegría o "zoale" (además de tamales y atole), se produce en pequeña escala en las tierras de temporal de las laderas cerriles y en las hoy resecas llanuras lacustres de los pueblos de Mízquic, Ixtayopan, Tecómitl, Tetelco, Tulyehualco y San Luis. Los dulces se elaboran en Tulyehualco sobre todo. Sus semillas se echan a nacer en abril, en almácigos de lodo en las chinampas (antes abonadas con guano de murciélago), de allí se trasplantaban las matitas a los suelos definitivos, cuando comienzan las lluvias. Es decir, se sigue un método semejante al observado para el chile, el cempasúchil y la lechuga, comentados anteriormente.

El terreno se prepara en forma similar al del maíz del cerro, con yunta y arado. Se colocan los chapines (que tienen de tres a diez plantas) a mano, uno en cada piquete de coa o pala, se tapan enseguida con pies y manos. Se da un primer fierro o escarda con arado para desyerbar y abonar (con estiércol o abono químico); más tarde se da el montón a cada mata, con azadón. La cosecha ocurre en diciembre o enero, cuando las panojas estan bien secas debido a que concluyó la temporada de lluvias y a la acción de las heladas. Por lo general la alegría se siembra sola, pero a veces se asocia con frijol de mata en el mismo surco, sembrado al mismo tiempo en grano. Otras veces se rota con frijol o bien con maíz, un ciclo y un ciclo.

Hoy en día el periodo agrícola del <u>uauhtli</u> o alegría - que fuera una de las plantas de mayor importancia alimenticia en la época prehispánica - es muy similar al del maíz chalqueño, pero ligeramente más largo (almácigo en abril, cosecha en diciembre). Esto quiere decir, si el ciclo no ha variado en forma notable con respecto al de aquella época, que la alegría para obtener semilla tampoco podía cultivarse dos veces al año, ni en tiempo de heladas. El hielo "mata la mazorca", la flor con sus semillas, me indicaron en San Luis, si bien la alegría resiste la sequía más que el maíz.

La cosecha de la alegría es un proceso complejo que requiere de gran habilidad y cuidado. Las ilustraciones del proceso que aparecen en el <u>Códice Florentino</u> (libro 4, capítulo 38), nos muestran que el de hoy es muy parecido al del siglo XVI. Ocurre de la siguiente manera: se cortan los tallos ya secos con hoz o machete, lo más bajo posible; se dejan en el suelo para que se sequen mejor al rayo del sol; se toman varias plantas para formar con ellas una "parva" encima de una manta. Se trillan primero con un garrote improvisado, luego con los pies, "bailando" encima de ellas para que se desprendan las semillitas de las flores. Con las manos se retiran las varitas, el resto se levanta con una palangana y se vacía en la "cernidera" que consisten en un ayate de ixtle amarrado a cuatro postes a determinada altura. La persona que cierne mueve la mano a manera de círculos, frotando sobre el ayate; la semillita blanca, de forma ligeramente arriñonada, va cayendo sobre un trapo limpio que se halla colocado abajo y el "tamo" (basura) se lo lleva el viento. La tarea requiere de maestría y de un día ventoso.

## <u>Plantas del Viejo Mundo</u>

Las especies vegetales del Viejo Mundo que fueron introducidas al país por los españoles desde principios de la época colonial, cayeron pronto en el terreno fértil de las chinampas y fueron incorporadas por los chinamperos a su lista de opciones agrícolas y comerciales. Las primeras que se mencionan son las lechugas ("Relación de Culhuacán", 1580) y unos "colinos" y "lechughuinos" (Ojea 1897:2-3), en el siglo XVI. Lechugas y coles estaban ya incorporadas como palabras al náuatl hacia 1570, según se desprende de su registro en el <u>Vocabulario</u> de fray Alonso de Molina, quien las traduce como <u>colestectli</u> y <u>lechugatectli</u>, <u>tepiton</u> <u>lechuga</u> y <u>uel</u> <u>quilechuga</u>. Un poco más tarde aparecen en las fuentes de la zona tanto las flores como las verduras introducidas. Si bien el número a finales del XVIII era similar al de los cultivos mesoamericanos (cfr. Alzate 1791, en el cuadro-resumen), aun no es posible establecer su importancia real en términos absolutos.

## Conclusiones

Para concluír este ensayo presento unas breves reflexiones en torno a las preguntas formuladas inicialmente.

Es una opinión más o menos generalizada que las plantas introducidas a México a partir de la conquista no sólo ampliaron y diversificaron el repertorio de las nativas, sino que, sobre todo los cereales, permitieron intensificar el uso del territorio, al poderse utilizar los terrenos durante el invierno y levantar dos cosechas al año (cfr. por ejemplo, Borah 1975:40). Respecto a las chinampas, más que considerar el papel jugado por los cereales, es pertinente examinar el complejo hortalizas-flores. La impresión inicial que se tiene al analizar este problema es que las plantas introducidas posibilitaron, al menos potencialmente, el aumento del número de cosechas en virtud de su capacidad para resistir los embates de las fuertes heladas.

Sin embargo las plantas del Viejo Mundo, adoptadas por los chinamperos a partir de la invasión colonial, representaron un papel más complejo que el de sólo "llenar" los tiempos de heladas invernales. Un primer efecto fue la suma y con ella la ampliación y diversificación del repertorio; la incorporación de especies de diversa naturaleza botánica, de distinto valor social y comercial, ofrecieron al productor mayores opciones y combinaciones, así como una mejor adaptación de sus medios de producción a las condiciones ambientales y a las demandas comerciales del mercado regional, especialmente el de la ciudad de México.

Las hortalizas mesoamericanas fueron los cultivos más importantes en el período de estudio, después del maíz, los cuales dominaron el paisaje agrario chinampero durante toda la colonia, el siglo XIX y hasta principios del XX. Asimismo considero, a nivel de hipótesis, que tuvieron más en este período que en el prehispánico tomando en cuenta, entre otras cosas, el aparente desplazamiento - en una época aun no precisada del periodo colonial - de las especies de <u>uauhtli</u> y chía empleadas principalmente como alimentos básicos en la elaboración de harinas y masa.

Estas hortalizas nativas no perdieron importancia sino hasta el siglo XIX. Ello ocurrió en algunas zonas durante las primeras décadas; en otras, hasta la cuarta o quinta. Hoy su cultivo ha desaparecido (como el del jitomate) o bien, es muy limitado (como el de chiles, calabaza, chilacayote o tomate).

Algunas hortalizas del Viejo Mundo se hicieron muy populares desde el siglo XVI, especialmente el pepino, la lechuga y la col, y se incorporaron plenamente a las chinampas, ampliando y diversificando el repertorio.

Pero fue en el transcurso de las primeras seis décadas del siglo XX, con cierta variación de un lugar a otro de las zonas chinamperas sobrevivientes, cuando tuvieron lugar los cambios más importantes del período de estudio. Las plantas nativas "comerciales" y las especies del Viejo Mundo que se fueron incorporando paulatinamente desde el siglo XVI (ya "tradicionales" en este sentido), fueron desplazadas casi por completo por plantas de ciclos más cortos así como por plantas de ornato, si bien muchas de ellas ya se cultivaban desde antes en pequeña escala.

Este cambio se debió a una combinación compleja de factores concatenados entre los cuales a veces resulta difícil distinguir los que fueron causa y los que fueron resultado. En primer lugar, cabe apuntar los profundos cambios ambientales ocurridos a partir de la intensa explotación de los acuíferos del sur de la Cuenca, especialmente los de la zona de Tláhuac y Xochimilco. Provocaron en forma más o menos directa el abandono de algunos cultivos "tradicionales" (nativos e introducidos) por falta de agua suficiente (jitomate, pepino, lechuga, tomate, etcétera) y la generalización de otros (hortalizas europeas y asiáticas así como nuevas plantas de ornato).

El crecimiento demográfico de la ciudad de México, que si bien fue la causa directa del aumento de la extracción de agua de las zonas circunvecinas a la ciudad, incluídas Xochimilco y otras, al mismo tiempo provocó otros efectos. Uno, que el área urbana se expandiera sobre antiguas zonas agrícolas; otro, que se crearan empleos no agrícolas en los servicios urbanos y la industria, ocupados principalmente por los campesinos de las zonas más cercanas, a menudo las mismas perjudicadas por la falta o la disminución de agua. El tercer efecto fue positivo, si bien efímero: la multiplicación de la demanda de productos agrícolas perecederos por parte de los habitantes de la ciudad de México. Resulta difícil situar con precisión las fechas de este período de aumento de la demanda de los productos de las chinampas, en gran medida porque no sucedió al mismo tiempo ni con la misma intensidad en los diversos pueblos chinamperos que todavía se mantenían. Sin embargo, la información indica que alcanzó su máximo entre 1940 y 1960. Fue éste el período en el que sobre todo pierden importancia las hortalizas mesoamericanas en tanto se intensifica la de las introducidas. Se impuso, como ya se dijo, la tendencia a producir plantas de ciclos más cortos, que son, por lo general, las de origen asiático y europeo así como plantas de ornato.

El aumento de la demanda urbana fue un poderoso e importante estímulo para la ampliación de las áreas de cultivo en las zonas chinamperas existentes, simultáneo al de su propia recuperación demográfica.

Pero tanto el crecimiento de la población como el desarrollo general del país, a partir de los cincuentas sobre todo, trajeron consigo la extensión de las redes de comunicación y de abastecimiento a la ciudad de México. Otras zonas productoras, más allá de la Cuenca de México como los Valles de Morelos, Cholula, Texmelucan, Tehuacán, etcétera, empezaron a competir con las de la propia Cuenca y terminaron por desplazar a muchos de sus productos tradicionales.

Table 8:1

## CUADRO RESUMEN

PLANTAS MENCIONADAS EN LAS FUENTES (Se respetó la forma de escribir de las fuentes)

Epoca Prehispánica

MAIZ, CHILE, TOMATE, CALABAZA, FRIJOL (Tezozomoc, 1944:17).

MAIZ, HUAUTLI, FRIJOL, CALABAZA, CHILCHOTL (ají o chile verde), JITOMATE. En camellones en Xaltocan (Tezozomoc, 1944:15).

MAIZ, FRIJOL, ALEGRIA, CHILE, CHIA, QUELITES, FLORES. Reconstrucción hecha por West y Armillas (1950: 180-181).

Epoca Colonial

1555 San Gregorio Atlapulco: JITOMATE (Anales de...:112)

1558 Xochimilco, jurisdicción: MAIZ en milpas de las tierra y del agua (Sobre el modo...:106)

1568 Xochimilco, pueblo: MAIZ, siembra en almácigo, estaba en fruto hacia mediados de marzo (A.G. Tierras, vol. 1525, exp.5).

1580 Culhuacan, jurisdicción: MAIZ, algún TRIGO, LECHUGAS, CALABAZAS, CHILES, TOMATES, QUELITES Y OTRAS LEGUMBRES E YERBAS QUE LOS NATURALES COMEN ("Relación de...", UTA XXIII-14).

1580 Mexicaltzingo, jurisdicción: MAIZ, QUILITES, CHILE Y OTRAS LEGUMBRES (PNE, VI:194).

1581 Ixtapalapa, señorío: TRIGO, CEBADA, FRIJOLES (Monjarás-Ruiz 1980).

1585 Xochimilco, jursidicción: MAIZ, CHILE, CHIA, (Ciudad Real, 1976, I:108).

1580 Culhuacan a Chalco: MAIZ, CHILE, COLINOS, LECHUGUINOS (Ojea 1897:2 y 3).

1599 Zona lacustre del valle en general: MAIZ Y OTRAS SEMILLAS (Vargas Machuca 1892, II:141-142).

1604 Zona lacustre en general: MAIZ, BERZAS (Torquemada 1969, II:483).

1698 Zona lacustre en general: MAIZ, LEGUMBRES, PIMIENTOS (chiles), TOMATES, XITOMATES (Vetancurt 1971:32-33). ZEMPOALXOCHITL O CLAVELLINA DE INDIAS (en particular en la ciudad de Xochimilco), MOSQUETAS, RETAMA, ESPUELA DE CABALLERO, CLAVELES (blancos, salpicados y colorados) (Vetancurt 1971:41 y 56).

1746 Xochimilco, ciudad: FLORES (Villaseñor 1952, I:165).

1746 Mexicaltzingo, jurisdicción: FLORES, HORTALIZAS (Villaseñor 1952, I:60).

1759 Iztacalco: VERDURA, comercio a Texcoco (AGN, Ríos y Acequias, vol. 4, exp. 1, f. 2r-3v).

1791 Santa Anita: MAIZ, HORTALIZA, FLORES, TOMATE, CHILE (AGN, Tierras, vol. 1220, exp. 1, f. 170).

1791 Culhuacán a Xochimilco: MAIZ, COL VERDULERO Y REPOLLO, BETUALLA:LECHUGA, ZANAHORIA, BETABEL, NABO, CHILE, TORNACHILE, TOMATE, JITOMATE, CALABAZA, PEREJIL, FLORES: SEMPOAXOCHITL, AMAPOLA, ALELI, CHICHARO DE CHINA,

Epoca Colonial (Cont.) Table 8:1 (cont)

| | |
|---|---|
| | ESPUELA DE CABALLERO (Alzate 1831, II:230-247). |
| ? | Xochimilco: FLORES, CEMPOALZOCHITL (Velázquez de León f. 74r-75r). |
| ? | " : VERDURA, VITUALLAS, FLORES (BNMadrid, 19266, f. 185v-185r). |
| 1803 | ¿Ixtacalco?: HABAS, GUISANTES, PIMIENTOS (Capsicum, chile), PATATAS, ALCACHOFAS, COLIFLOR, "E INFINIDAD DE OTRAS VARIAS LEGUMBRES" (Humboldt 1966:135). |

Epoca Independiente y Porfiriato

| | |
|---|---|
| 1816 | Xochimilco a Ixtayopan: FLORES, VITUALLAS, algún MAIZ; 4 años de anegación (AGN, Desagüe, vol. 43, exp. 5, no. 465). |
| 1826 | Iztacalco, municipio (Santa Anita, San Juan Nexticpac, La Magdalena Atlazolpa): TORNACHILE, CHILITO, TOMATE, JITOMATE, CALABAZA, LECHUGA: dañadas por una granizada el 24 de mayo (AGN, Ayuntamientos, vol. 10, f. 237r-274v). |
| 1855 | Santa Anita e Iztacalco: HORTALIZA, FLORES: ROSAS DE CASTILLA, CLAVELES, AZUCENAS, AMAPOLAS, CHICHAROS (Payno 1855:412-413). |
| 1864 | Chalco: FRUTAS, LEGUMBRES (Orozco y Berra 1864:165). |
| 1883 | Iztacalco, Santa Anita, Los Reyes: AMAPOLAS (rosas y blancas), ACIANOS AZULES, CHICHAROS MORADOS, ALHELIES, ESPUELINES, CAMPANULAS, TREBOL, MASTRANZO (Altamirano 1949:210). |
| 1883 | Tetelco: COLES (Peñafiel 1884:83). |
| 1884 | Zona lacustre en general: FLORES Y LEGUMBRES (García Cubas 1884:262). |
| 1894 | Santa Anita, Iztacalco, San Juanico, Mexicaltzingo e Iztapalapa: AMAPOLAS, ESPUELA DE CABALLERO, CLAVELES, ZEMPAXUCHIL, LEGUMBRES Y HORTALIZAS (García Cubas 1894:19). |
| 1894 | Iztapalapa: MAIZ (además de las anteriores) (García Cubas 1894:19). |
| 1899 | En la llanura del canal (de la Viga): VERDURAS (toda clase), y algunas LEGUMBRES Y FLORES. (Figueroa D. 1899, I:356). |
| 1904 | Santa Anita e Iztacalco: HORTALIZA, LECHUGAS, APIO, AMAPOLAS, ESPUELA DE CABALLERO, CLAVEL, RETAMA, CHICHAROS, ALHELIES, ZEMPOALXOCHITL (García Cubas 1904:323). |
| 1911 | Xochimilco: MAIZ, JITOMATE, CHILE, COL, COLIFLOR, LECHUGA, TOMATE, COL DE BRUSELAS, CEBOLLA, ESPINACA, APIO, RABANO, AMAPOLA, MARGARITA, CLAVEL, PENSAMIENTO, CRISANTEMA, DALIA, ALELI (Santamaría 1912:14-15 y ss.). |
| 1911 | Iztapalapa: ALCACHOFA, AJO, COL, CHILE, ACELGA (Santamaría 1912:14-15 y ss.). |
| 1911 | Iztacalco, Santa Anita: LECHUGA, CEBOLLA, PORO O PUERRO, NABO, ZANAHORIA, AMAPOLA, MARGARITA, CLAVEL, PENSAMIENTO, CRISANTEMA, DALIA, ALELI (Santamaría 1912: 14-15 y ss.). |

Table 8:1 (cont)

Epoca Independiente y Porfiriato

1911 Zona chinampera Xochimilco a Iztacalco, cultivos de menor importancia: ACELGA, ACEDERA, PEREJIL, CULANTRO, RABANO, CALABAZA, FRIJOL, PEPINO, APIO, ALFALFA, CEBADA (Santamaría 1912:14-15 y ss.).

Epoca Post-Revolucionaria

1923 Zona chinampera en general: MAIZ, FLORES, VEGETALES (Nuttall 1925:463).

1926 Mexicaltzingo, Iztacalco, Santa Anita: MAIZ, JITOMATE, CHILE, CALABACITAS, TOMATES, LECHUGAS, COLIFLOR, CEBOLLA, APIO, ESPINACA, ALCACHOFA, AJO, PUERRO, NABO, ZANAHORIA, ACELGA, ACEDERA, PEREJIL, CULANTRO, RABANO, PEPINO, ALFALFA, CEBADA; AMAPOLA, CLAVEL, MARGARITA, CRISANTEMA, PENSAMIENTO, DALIA, ALELI, ALCATRAS (Montes de Oca 1926:424).

1938 Zona chinampera de Xochimilco en general: MAIZ Y FRIJOL (autoconsumo); COL, ESPINACA, LECHUGA, PEPINO, APIO, JITOMATE, CHICHARO, EJOTE, CEBOLLA, LIRIOS, CLAVEL, AMAPOLA, TULIPAN, NOMEOLVIDES, ALCATRAZ, NARDO, CRISANTEMO, PENSAMIENTO, MARGARITA (Schilling 1938).

1946-1947 Mixquic: MAIZ, TOMATE, JITOMATE, COLIFLOR, CILANTRO, ALEGRIA, CHILE, ZANAHORIA, NABO, RABANO, CEBOLLA (West y Armillas 1950; Armillas, notas 1946).

Ixtacalco: MAIZ, LECHUGA, COL, FRIJOL, PORRO, CEBOLLA, ZANAHORIA, ALCACHOFA, CILANTRO, BETABEL, ALFALFA, TOMATE, JITOMATE, NABO, IMPERIAL, CLAVEL, CHICHARO, GLADIOLA, MALVA (West, notas 1947).

Ixtapalapa: LECHUGA, BETABEL (West, notas 1947).

1953-1954 MAIZ ALEGRIA, JITOMATE (Sanders 1957).

Table 8:2

LISTA DE ESPECIES*

| | |
|---|---|
| ACELGA | *Beta vulgaris* var. *cicla* L. |
| AGAZANIA (GAZANIA) | *Gazania* spp. |
| AJO | *Allium sativum* L. |
| AHUEJOTE | *Salix bomplandiana* H.B.K. |
| ALBAHACA | *Ocimum basilicum* |
| ALCACHOFA | *Cynara scolymus* L. |
| ALCATRAZ | *Zantedeschia* spp. |
| ALEGRIA O UAUHTLI | *Amaranthus leucocarpus* S.Wats |
| ALFOMBRILLA | *Verbena pulchella* |
| ALHELI | *Matthiola incana* (L.) Br. et R. |
| ALVERJON | *Pisum arvense* L. |
| AMAPOLA | *Papaver rhoeas* o *P. somniferum* L. |
| AMAPOLA JAPONESA | *Eschscholtzia californica* |
| APIO | *Apium graveolens* L. |
| ARETILLO | *Fuchsia fulgens* |
| AZUCENA | *Lilium candidum* L. |
| BETABEL | *Beta vulgaris* var. *crassa* Alef. |
| BROCOLI | *Brassica oleracea* var. *cauliflora* |
| CALABAZAS: | |
| a) calabacita japonesa | *Cucurbita pepo* L. |
| b) calabacita redonda | *Cucurbita pepo* L. |
| c) calabaza de Castilla | *Cucurbita mixta* Pang. o *C. moschata* Duech. |
| CAPULIN | *Prunus capuli* C. |
| CEBADA | *Hordeum vulgare* L. |
| CEBOLLA | *Allium cepa* L. |
| CILANTRO | *Coriandrum sativum* L. |
| CIRUELO | *Prunnus* spp. L. |
| CLAVEL | *Dianthus caryophyllus* |
| COL | *Brassica oleracea* var. *capitata* L. |
| COL DE BRUSELAS | *Brassica oleracea* var. *gemmifera* L. |
| COLINABO | *Brassica oleracea* var. *napobrassica* Mill. |
| COLIFLOR | *Brassica oleracea* var. *botrytis* L. |
| CRISANTEMO | *Chrysanthenmum* spp. L. |
| CHABACANO | *Prunus armeniaca* L. |
| CHAYOTE | *Sechium edule* (Jacq.) Sw. |
| CHIAS | *Salvia hispanica* L. o *Hyptis suaveolens* Pot. |
| CHICHARO DE OLOR | *Lathyrus odoratus* L. |
| CHILACAYOTE | *Cucurbita ficifolia* Couche |
| CHILES | *Capsicum annum* L. y *C. frutescens* L. |
| CHINO | *Impatiens sultani*, *I. holstii* |
| DALIA | *Dahlia coccinea* Cav., *D. pinnata* Cav., *D. lehmannii* Hieron |
| DURAZNO | *Prunus persica* Stokes |
| EPAZOTE | *Chenopodium ambrosioides* |
| ESPINACA | *Spinacia oleracea* L. |
| ESPUELA DE CABALLERO | *Delphinium ajacis* |
| ESTER (ASTER) | *Aster* spp. |
| FLOR DE MUERTO O CEMPASUCHIL | *Tagetes patula* L., *T. erecta* L. |
| FRIJOLES: | |
| a) delgado | *Phaseolus vulgaris* L. |
| b) gordo o ayecote | *Phaseolus coccineus* L. |
| GLADIOLA | *Gladiolus* spp. |

*La lista de la izquierda es la de los nombres vulgares utilizados en las fuentes y en la actualidad en las zonas chinamperas. La de la derecha es la de los nombres científicos, sacada de diversas fuentes.

| | |
|---|---|
| GRANADA | *Punica granatum* |
| HABA | *Vicia faba* L. |
| HIGO | *Ficus carica* |
| IMPERIAL | *Centaurea imperialis* |
| JITOMATE | *Lycopericon esculentum* Mill. |
| JUANITA | *Phlox drummondii*, P. *subulata*, P. *suffruticosa* |
| LECHUGA | *Lactuca sativa* L. |
| LENTEJA | *Lens culinaris* |
| LIRIO | *Iris spuria* |
| MAGUEY | *Agave atrovirens* Karw., A. *latissima* Jacobi, A. *pisaga* Trel. |
| MAIZ | *Zea mays* L. |
| MALVA | *Malva* spp. |
| MANZANILLA | *Matricaria chamomilla* |
| MARGARITA | *Callistephus chinensis* (L.) Nees |
| MEMBRILLO | *Cydonia oblonga* Mill. |
| MERCADELA | *Calendula officinalis* L. |
| NABO | *Brassica napus* |
| NOMEOLVIDES | *Myosotis sylvatica* |
| NOPAL | *Opuntia* spp. |
| NUBE | *Gypsophila paniculata* L., G. *elegans* |
| OLIVO | *Olea europaea* |
| PANALITO | *Alyssum maritumum* |
| PENSAMIENTO | *Viola tricolor* L. var. *hortensis* |
| PEPINO | *Cucumis sativum* L. |
| PEREJIL | *Petroselinum crispum* Hoffm. |
| PETUNIA | *Petunia hybrida* |
| PINCEL | *Centaurea cyanus* |
| PORO O PUERRO | *Allium porrum* L. |
| QUELITES | *Amaranthus* spp., *Chenopodium* spp., *Porophyllum* spp., *Rumex* spp. |
| RABANO | *Raphanus sativus* L. |
| ROMERITO | *Suaeda diffusa* Wats. |
| ROMERO | *Rosmarinus officinalis* L. |
| RUDA | *Ruta graveolens* L. |
| SALSIFI | *Tragopogon porrifolius* |
| TEJOCOTE | *Crataegus pubescens* (H.B.K.) Stead. |
| TOMATE | *Physalis ixocarpa* Brot. |
| TRIGO | *Triticum sativum* Lam. |
| TUBEROSA | *Begonia tuberosa* |
| UAUHZONTLE O QUELITE MAZORCA | *Chenopodium nuttaliae* Saff. |
| VARA DE SAN JOSE | *Althaea rosea* |
| VIOLETA | *Viola odorata* L. |
| YERBABUENA | *Mentha sativa* L. |
| ZANAHORIA | *Daucus carota* |

Table 8:2 (cont)

AGRADECIMIENTOS

La investigación documental y de campo necesarias para realizar este ensayo fue posible gracias al apoyo que me brindó el CISINAH (hoy CIESAS), de 1976 a 1979. Quiero dejar constancia de mi gratitud a los chinamperos de las zonas de Xochimilco, Mízquic e Itztapalapa por su paciencia para enseñarme sus complejos conocimientos, en especial a los de San Luis de Tlaxialtemalco, Xochimilco. A los profesores Pedro Armillas y William T. Sanders por lo que de ellos aprendí durante el Primer Taller de Arqueología Superior, celebrado en 1973. Nuevamente mi agradecimiento a Pedro Armillas por haberme facilitado una copia de sus notas inéditas y otra de las de Robert C. West. A los maestros Jesús Ruvalcaba y Luis Reyes, por sus útiles comentarios a este escrito. A Lucy Pereyra por la corrección del texto y a José Luis Ramos por realizar los dibujos que lo ilustran.

Chapter 9

# PRELIMINARY RESULTS OF THE SAMBORONDON RAISED FIELD PROJECT, GUAYAS BASIN, ECUADOR

William M. Denevan and Kent Mathewson

Abstract

A large collection of raised fields in the Middle Guayas Basin, north of Guayaquil, were first documented in the last century. Some were mapped and excavated during the 1979-80 Samborondón project, the results of which are reported here. In the Guayas Basin ceremonial mounds, or tolas, habitation mounds, or lomas, and raised fields extend over hundreds of hectares. Twenty-three sites were located and tested and four raised field zones were identified. All but one of the sites showed a predominance of Milagro-Quevedo pottery. It is unlikely that the Samborondon area was the heart of raised field agriculture but early $C^{14}$ dates derived from soil beneath raised fields yields dates of ca. 2000-2500 B.C., the earliest dates yet to be associated with drained fields. On the basis of form and function nine distinct complexes of raised fields were distinguished in the Guayas Basin. Lithic and ceramic evidence from the surrounding habitation mounds suggests that maize and manioc were grown on the raised fields. Excavation of the mounds allows an easy identification of their use as the difference in artifact density between habitation mounds and raised fields is 500:1. Estimates of population density which could have been supported by the drained fields are made.

Resumen

Una gran colección de campos elevados en la cuenca del Guayas medio, al norte de Guayaquil, fue documentada por primera vez en el siglo pasado. Algunos fueron trazados en el mapa y excavados durante el proyecto Samborondón 1979-80, cuyos resultados se presentan aquí. En la cuenca del Guayas, los terraplenes ceremoniales o tolas, terraplenes habitacionales o lomas y los campos elevados cubren cientos de hectáreas. Se ubicaron y se efectuaron pruebas en veintitrés sitios y se identificaron cuatro zonas de campos elevados. En todos los sitios, con la excepción de uno, dominó la cerámica de Milagro-Quevedo. No es probable que el area de Samborondón haya sido la cuna de la agricultura de campos elevados pero fechas de $C^{14}$ derivadas del suelo debajo de los campos elevados indican fechas de alrededor de 2000-2500 A.C. que constituyen las fechas más antiguas hasta ahora que se asocien con campos drenados. Sobre la base de forma y función se distinguieron nueve complejos distintos de campos elevados en la cuenca del Guayas. Evidencia lítica y de cerámica de las lomas vecinas sugiere que se cultivaron el maíz y la mandioca en los campos elevados. La excavación de las lomas permite una fácil identificación de su uso ya que la diferencia en la densidad de artefactos entre las lomas y los campos elevados es de 500:1. Se calculan densidades de la población que los campos drenados podrían haber sostenido.

## Introduction

Remnants of pre-Columbian raised fields have now been discovered by geographers and archaeologists in many parts of South America and Mesoamerica (Denevan, 1982). These are ridges or platforms of variable height, width, and length whose primary function seems to have been drainage in order to improve conditions for cultivation. They are found in both lowlands and highlands in swamps, lakes, and periodically flooded basins and river valleys. Most were abandoned before or shortly after the Spanish Conquest.

One of the largest concentrations of raised fields is in the Middle Guayas Basin east of the city of Guayaquil in coastal Ecuador (Figures 9:1, 9:2). Here there are large savannas which are periodically flooded by the overflowing of the Guayas, Babahoyo, Daule, and other rivers rising in the Andes. The area is currently undergoing mechanised wet-rice development, and the old fields and associated earthworks are rapidly being destroyed and with them the opportunity to study the evolution and ecology of intensive agriculture in a lowland floodplain. Raised fields have also been found in small clusters in several highland basins of Ecuador, including Cayambe, Quito, and San Pablo; (see Batchelor 1980; Knapp and Ryder n.d.; and Knapp 1983.)

This report reviews previous research and presents the preliminary results of the 1979-1980 Samborondón Project. A new map (Figure 9:2) is provided of the distribution of the relic raised fields and of morphologically differentiated field complexes in the Guayas Basin. Some suggestions are made regarding population, age, and origins of the fields.

## Landscape Archaeology of the Guayas Basin in Perspective

Since the earliest days of the Conquest, it has been generally known that prehistoric Guayas Indians moved large amounts of earth to make the ceremonial mounds or *tolas* that occur by the thousands in much of coastal and highland Ecuador. Some of these *tolas* in the areas of raised fields are as much as 12 metres high. Speculation concerning their origin centred on either the local Huancavilcas or the Incas. Among the first archaeological attempts to investigate the Guayas earthworks was Otto von Buchwald's (1917) excavation of a *tola* on Hacienda Chilintomo. Theodore Wolf (1892: 118), the great *fin de siecle* natural historian of Ecuador, was quite meticulous in classifying 13 different biotopes found in the coastal savanna zone, but he failed to realise that most of the islets dotting the swamps were artificial. He did note, however, that the depressions were slowly being silted in by annual inundations.

Modern archaeological study of the Guayas aboriginal earthworks began in the 1950's with the Guayaquileno archaeologist Emilo Estrada along with Betty Meggers and

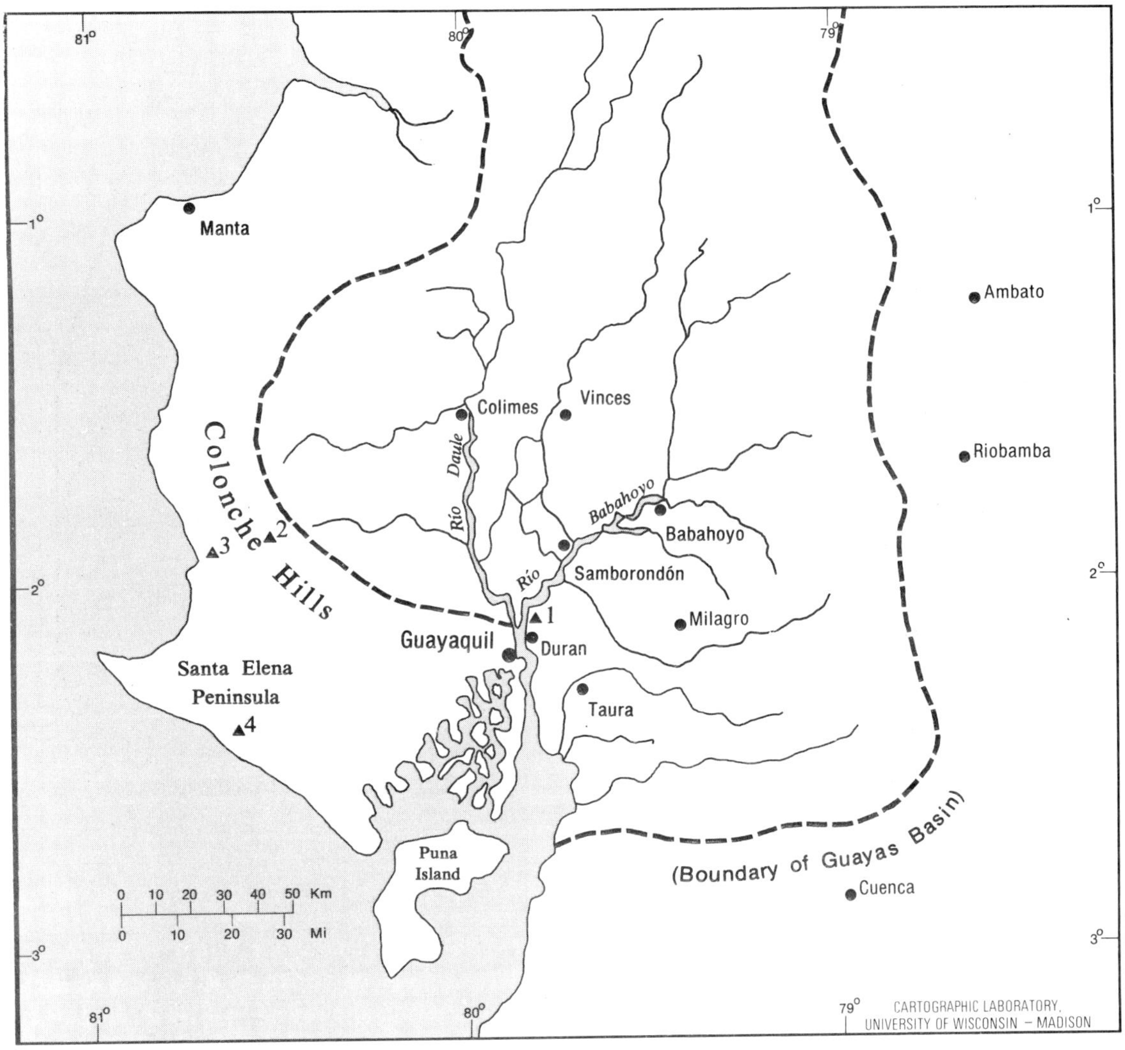


Figure 9:1 The Guayas Basin. Archaeological sites: 1 Peñon del Río; 2 Loma Alta; 3 Valdivia; 4 Real Alto

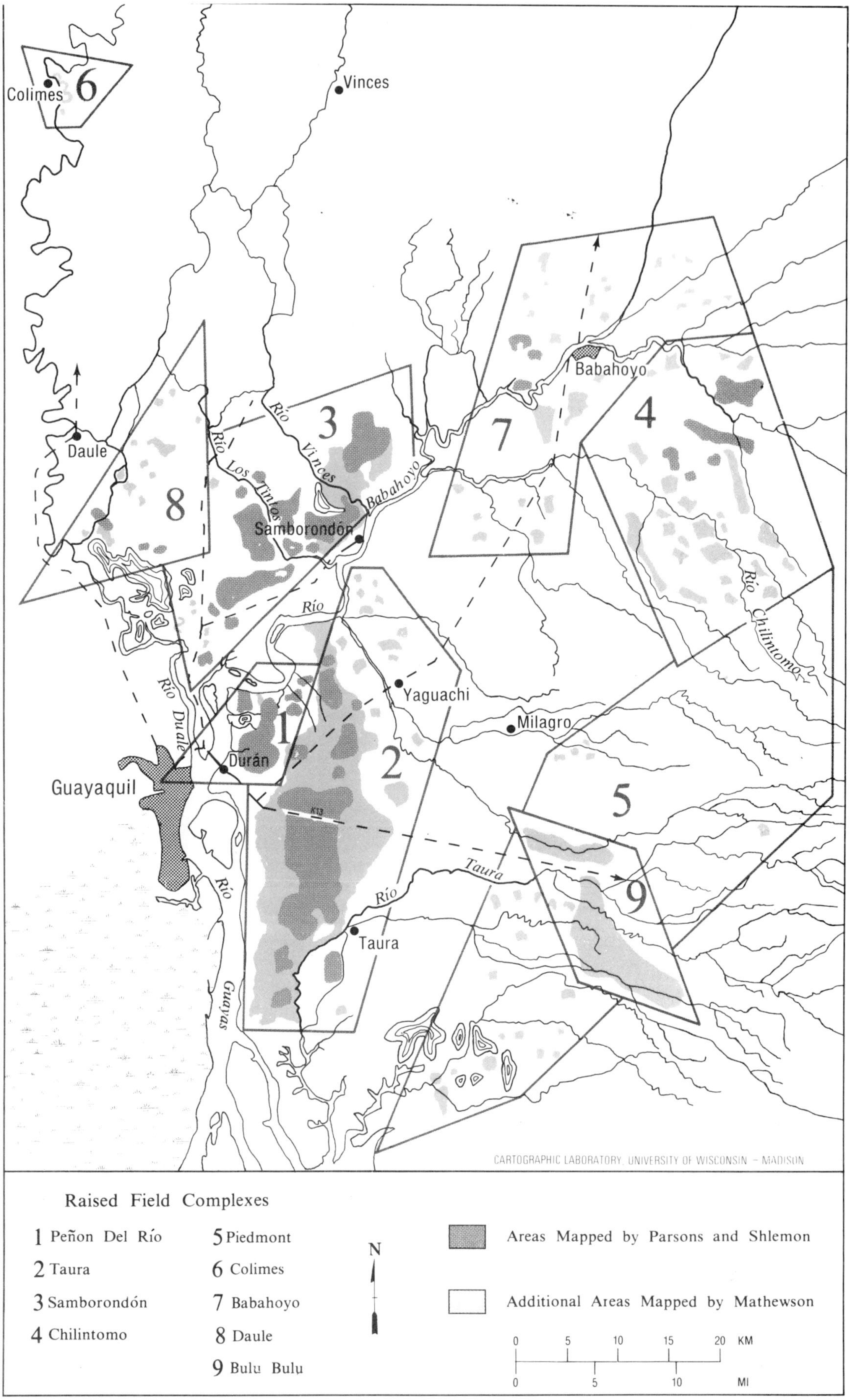


Figure 9:2 Raised field complexes of the Guayas Basin

Clifford Evans (Estrada 1957). In summarising their work, Meggers (1966: 132) suggested that the artificial mounds that "occur by the hundreds...in the southern Guayas basin... where the land is inundated every year for several months during the rainy season," were built for habitation sites or for ceremonial and/or funerary purposes, but an agronomic function was not considered. She associated the mound building with the Milagro cultural phase that occurred during the last millenium before conquest.

In 1965 Berkeley geographer James J. Parsons first recognised raised fields while flying into the city of Guayaquil. Subsequent reconnaissance from a small airplane and on the ground confirmed the location of three different complexes. Parsons (1969: 78) estimated that the group closest to Guayaquil, the Peñon del Río Comlex of linear, ridged raised fields, encompassed at least 1,600 hectares of ridges and swales. Parson's recognition of the agro-economic nature of the thousands of mound, ridge, and platform structures that occur throughout the Lower Guayas Basin came after similar complexes had been discovered just a few years before in Bolivia (Denevan 1966) and Colombia (Parsons and Bowen 1966). Parsons returned to Guayas in 1976 with geomorphologist Roy Shlemon to further map the fields and to attempt dating (Parsons and Shlemon 1982).

## Excavations at Samborondón

At the beginning (June-August) and at the end (December) of the 1979 dry season the University of Wisconsin carried out an archaeological survey in a 25 square kilometre area near the town of Samborondōn on the Río Babahoyo northeast of Guayquil (Figure 9:1). From June 1979, to December 1980, a cultural geographic study was made of different biotope use throughout the cantōn of Samborondón and in other raised field districts in the Guayas basin (Mathewson 1980). At Samborondōn 23 sites were located and tested; in addition, four relic raised-field zones were identified. In zones I and II, one pair each of rectangular platform fields (<u>bancos</u>) and their intervening ditches were excavated. In zones I, II and III the <u>lomas</u> or habitation mounds closest to the excavated fields were also tested.

While none of the raised fields excavated by the Samborondon Project yielded enough carbon for dating, two lomas did provide datable quantities of carbon (Bender <u>et al</u>. 1981: 150). Samples taken from an ash lense in Loma Desirée date to 1270 ± 70 B.P. or about A.D. 710 (WIS Radiocarbon No.1150). Bones taken from a collapsed burial urn in Loma del Diablo date to 1350 ± 70 B.P. or about A.D. 630 (WIS Radiocarbon No.1145). These dates suggest minimal age of the mounds, particularly in the case of the burial urns which were almost certainly intrusive (Olaf Holm, personal communication). The dates suggest an early Milagro-Quevedo phase (A.D. 500 to 1500) occupation.

Assuming that the habitation mounds were built coeval with the adjacent raised fields, then the structures were probably a simple manifestation of the population expansion generally ascribed to the late Integration Period which began around A.D. 500 and ended with the Spanish Conquest in 1535 (Meggers 1966: 25). However, it should be stressed that these dates come from only one small area of the Samborondón Complex, and from only one of the several raised-field complexes.

The ceramic evidence indicates an absolute predominance of Milagro-Quevedo phase pottery in all of the sites tested save one. This was a large 2.5 metre high *loma* construction that probably constituted the aboriginal site of Samborondón. Donald Lathrap, Jorge Marcos, and Louisa Stark surface collected the site in 1977. Within a thick deposit of Milagro-Quevedo phase material, they found a single sherd thought to be Valdiva phase pottery. The mound was bulldozed in 1979 to make a soccer field, and in the remains Mathewson found Guangala phase (500 B.C. - A.D. 500) polychrome sherds and a copper or bronze axe head. Otherwise, the sites tested in the Samborondón area have uniformly late pottery remains. They express little variation in rim style or design. From the extremely unrefined nature of the material recovered, it seems that the people living on or near the ancient raised fields had little power or desire to consume elaborately made goods. While there was a general diminution in ceramic quality and execution after A.D. 500 throughout ancient Ecuador, often attributed to increased mass production, the Milagro-Quevedo artefacts collected in context with the Guayas raised fields are particularly uninspired. This suggests that if surplus wealth was generated by raised-field farming, its material expressions were deposited elsewhere, most notably outside the immediate raised-field districts in *tolas* containing gold figurines and other objects of elite culture of the Milagro-Quevedo phase.

The evidence from Samborondón suggests that both maize and manioc were grown on the raised fields. While we lack pollen evidence, the lithic and ceramic material recovered supports this possibility. From the earliest Formative times, flint corns would have been available to the Guayas farmers (Lathrap *et al*. 1975: 20). For preparation of large kernel corn, stone metates are useful, and metates were surface collected on several *lomas* in the Samborondón area. The evidence for manioc comes from the large quantity of ceramic, grater-bowl sherds found. These are thought to be indicative of root-crop preparation. Similarly, numerous obsidian bladelets were uncovered in the excavations. Lathrap *et al*. (1975: 21) have suggested that such tools set in wood were used for manioc scrapers. Obsidian was a trade item. It does not occur in the Guayas floodplain, and probably came from the Andean highlands. Today, both maize and manioc are grown on relic raised fields and on recently reconstructed rice-field embankments; however, the old fields are not repaired or rebuilt. Maize and manioc are sometimes intercropped. Other staples may have been peanuts, sweet potatoes, mangareto (Xanthosoma), arrowroot, and the New World yam (Dioscorea trifida), all subsurface producers. Manioc, peanuts, and sweet potatoes did well in our one year experimental raised-field plot at Samborondón, until destroyed by cattle.

To generalise on the origins of this farming system for the basin as a whole, on the basis of the Samborondón sample, seems ill-advised. There is not enough environmental uniformity between the locations where relic raised fields occur or morphological similarity between complexes to argue for simultaneous origins. It seems that the fields we tested were part of the Integration Period changes that transformed large sectors of the Guayas landscape. However, it would be misleading to assume that this process developed here originally without strong antecedence, both locally and in other sectors of the Guayas Basin. For example, the fields tested at Samborondón are within a few hundred metres of the main river, the Babahoyo. In prehistoric times these fields were probably closer to the river judging from changes in the river channel as viewed from aerial photographs and on the ground. If so, then they were in the back swamp immediately behind the natural levee. They probably represent an effort to reproduce and expand the levee environment.

The Samborondón subarea, although rich in evidence for raised-field farming, with some outstanding ensembles of well articulated field patterns, was probably not the hearth of this intensive agro-adaptation. The absence of pre-Milagro-Quevedo cultural associations, and the close concurrence of the two carbon dates with each other and with the ceramic age, suggest that raised-field farming came to the Samborondón area as an already coherent technological and ideological entity. It would be worthwhile to look elsewhere in the Guayas basin, and perhaps beyond, for the antecedents of the Samborondón raised fields.

## Raised-field Complexes and Area Covered

We have identified and named nine distinct complexes of raised fields in the Guayas Basin, each with markedly different patterns. These are mapped on Figure 9:2 and are illustrated in Figures 9:3 and 9:4. Briefly, these complexes are differentiated as follows:

1. Peñon del Río - long, linear ridged fields, with general north-south orientations.

2. Taura - medium sized, circular to retangular platform fields, with some bundles of sub-parallel, linear ridged fields.

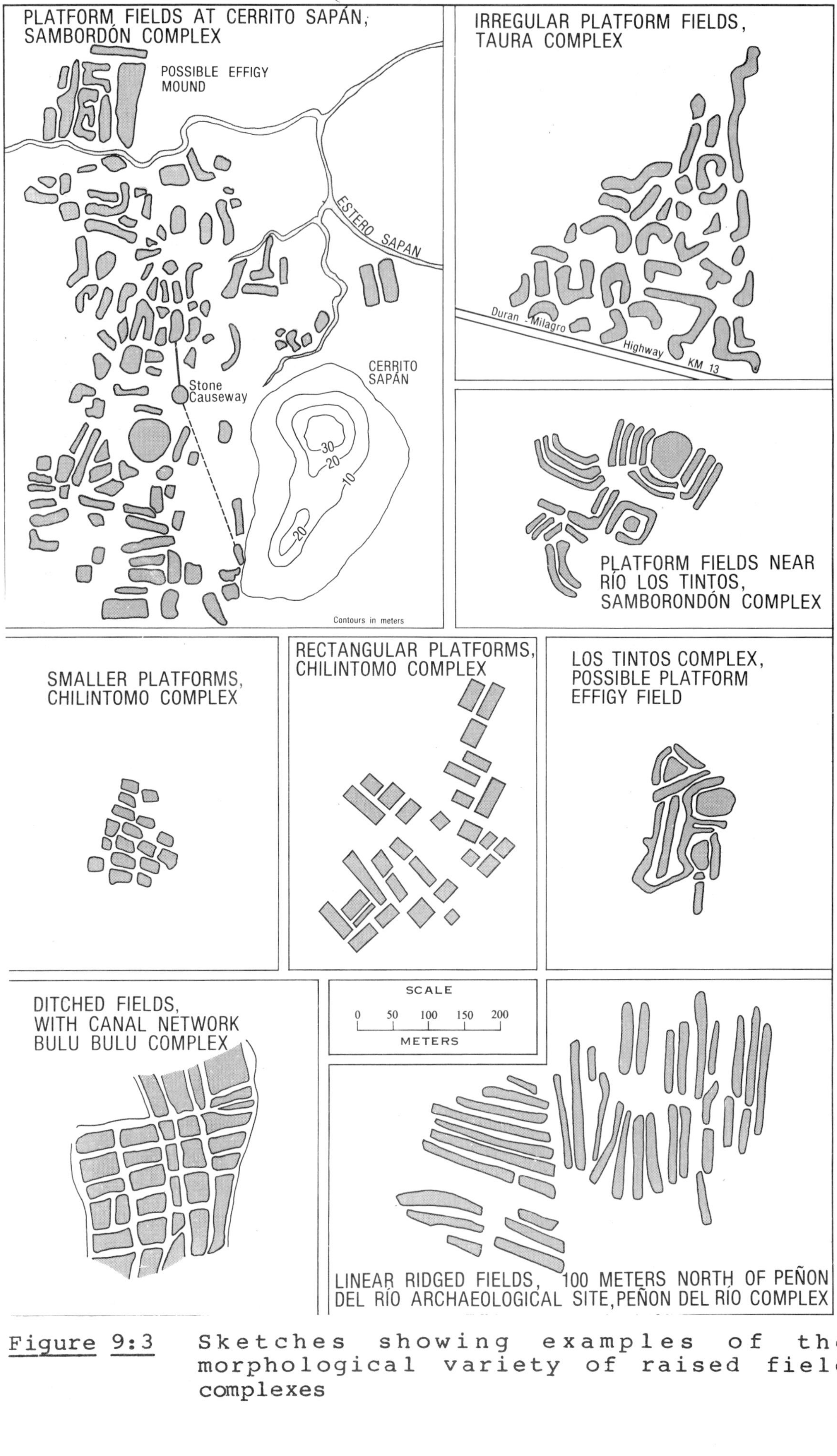


<u>Figure</u> <u>9:3</u> Sketches showing examples of the morphological variety of raised field complexes

Figure 9:4 Air photograph showing raised field patterns in the Samborondon complex along the Rio Tinto. Scale ca. 1:40,000, August, 1966

3. Samborondón - bundles of short, linear ridged fields interspersed with mounds and rectangular platform fields.

4. Chilintomo - extremely regular rectangular platforms, possibly for specialised arboriculture.

5. Piedmont - irregular platforms, along the middle courses of rivers originating in the Andes.

6. Colimes - tight checkerboard, parallel platforms within oxbows.

7. Babahoyo - irregular to rectangular platforms, with some checkerboard patterns.

8. Daule - irregular to rectangular platforms, mostly in abandoned river channels.

9. Bulu Bulu - apparent ditched fields, with little or no raised surface.

Three of these complexes were identified and partially mapped by Parsons (1969), in 1965 (Peñon del Río, Taura and Samborondón), and in 1975 he and Shlemon carried out additional field and air photo research. They mapped the location of additional fields, including part of the Chilentomo complex, for a total of 24,000 hectares of field and ditch surface (Parsons and Shlemon n.d.). Based on a more extensive examination of air photos, Mathewson has added five more complexes and expanded the total area of field and ditch surface to a minimum of 50,000 hectares. The photos used were taken in 1966, and many of the fields visible then have since been destroyed by mechanised rice cultivation. Many fields were buried long ago by natural siltation. (Lathrap personal communication, has suggested, given the alluvial history of the region, that if the earliest fields were built in the Formative Period there may even be strata of raised fields buried beneath the currently visible features.) Numerous surviving fields are obscured by forest or water, and are not readily visible from the air. Also, some of the earthworks are round to oval or irregular shaped mounds and are not distinguished at first glance as raised fields. These structures may have been built first as habitation mounds and later used as raised fields or vice versa. Distinguishing between relic habitation mounds and raised fields presents a problem in several of the Guayas complexes. From the Samborondon ceramic material, Richard Whitten (personal communication) estimates that "the difference in artefact density (between habitation mounds and associated raised fields) is on the order of 500:1." Thus, a distinction can be made through excavation. Generally, estimates of area of fields are difficult, and the figure of 50,000 hectares is conservative.

Some Speculative Conclusions

Population. Some speculation about the size of the population supported by the Guayas raised fields is possible by calculating the area of raised-field surface. The ratio of actual field surface to the total area of each complex varies widely. Rarely does it exceed a ratio of fifty percent field surface to intervening space, nor does it fall much below ten percent. Using a twenty percent overall ratio, based on sampling from air photos and ground measurements, and using a 50,000 hectare total, we derive a rough estimate of 10,000 hectares of raised-field surface.

If the figure of 19 persons per hectare (Sanders et al. 1979: 380) for Mexican chinampa cultivation (maize, one crop/year) can be transferred to the Guayas, and if we can postulate an unlikely period in which all fields were under cultivation simultaneously, then the number of people that the Guayas raised fields could have sustained with a maize staple was 190,000; the total would be higher with two crops a year, but lower if all fields were not cultivated at one time. If the staple was manioc and using an average of 16 people supported per hectare, the total population would have been 160,000 (Denevan 1982: 190). This is not a maximum prehistoric population figure for the Lower Guayas basin however. Other micro-environments would have been cultivable, and probably were. For example, today shifting cultivation is carried out on the large hillocks that rise like oceanic islands above the swamps and savannas in several of the raised-field complexes. The natural river levees are usually cultivated too. Nor should we assume that the Guayas raised fields were used only for subsistence purposes. Today cotton and cacao are grown on relic raised fields because they provide dry surfaces during flooding. Some commercial specialisation and monocropping also may have been done in ancient times.

Whatever the population, the aggregate area of moved earth in the raised fields of the Guayas Basin was enormous, possibly 100 million cubic metres requiring 38.5 million labour days (Denevan 1982: 190).

Age. In order to make demographic projections from relic raised fields, we must know much more about the antiquity and length of construction periods of different complexes. In addition to the dating already described for Samborondón (ca. A.D. 605 and 685), Parsons and Shlemon (n.d.) retrieved dates from paleosols at the base of raised fields in the Taura and the Peñon del Río Complexes. From a buried vertisol in a swale between two ridged fields at Cano Matanzas in the Peñon del Río Complex, they obtained a date of 3995 ± 95 years B.P. (ca. 2005 B.C.); and from the mollic epipedon at the base of an elliptical small platform field near Taura they obtained a date of 2450 ± 85 years B.P. (*ca.* 590 B.C.) (Parsons and Shlemon n.d.). So far, the only other remotely comparable date for a New World raised field comes from Puleston's (1977: 452) radiocarbon dating of a worked timber found in a canal at the Rio Hondo Complex in Belize (1110 ± 230 B.C.). Other datings of raised fields elsewhere suggest they are younger by a millenium or more (Turner and Harrison 1981: 402; Zucchi and Denevan 1979: 642; Armillas 1971: 658; Bommert 1976: 138; Siemens and Puleston 1972: 234). The Parsons-Shlemon early dates for buried soils need to be confirmed by ceramic associations of comparable ages.

*Origins*. The above soil dates are by no means conclusive, but they raise the possibility of Formative period (3500 B.C. to 500 B.C.) raised fields in Ecuador. But until equally early cultural remains are recovered in clear context with raised fields, the notion of the Guayas as a hearth of South American raised fields must be viewed with considerable caution. It is precisely because of the extreme cultural precocity of coastal Ecuador that we must be careful. A mangrove/maritime variant of Sauer's (1948) "mesolithic progressive fisherfolk - horticulturalists" seem to have been settled and farming by 7000 B.C. on the Santa Elena Peninsula of Ecuador (Stothert n.d.). The earliest New World evidence for ceramic making comes from the Colonche Hills piedmont ecotone between the coastal edge and the Guayas savannas (Norton n.d.: Lathrap *et al.* 1975: 15). The spatial evidence for social differentiation and the building of a ceremonial centre by 3500 B.C., the earliest in South America, has been uncovered by Lathrap, Marcos and Ziedler (1977) at Real Alto near the coast west of Guayaquil. Finally, the Guayas basin is a prime candidate as a locus of early plant domestication (Damp *et al.* 1981), as well as intensive maize agriculture (Zevallos *et al.* 1977).

During the 1980 field season, the authors and Julio Viteri discovered the first Valdivia occupation to be uncovered in the eastern Guayas Basin. The site, called Peñon del Río after an adjacent hillock, is at the centre of the Peñon del Río Complex of raised fields. Excavation of the Peñon del Río site is in progress under the direction of Dr. Jorge Marcos, Director of the new School of Archaeology at Guayaquil's Polytechnic Institute (Marcos 1982). The site promises to yield stratigraphic evidence of continuous occupation from at least Middle Valdivia (*ca.* 2,500 B.C.) to the time of European Contact. The large site is within one hundred metres of a group of well articulated subparallel ridged fields and is three kilometres north of the fields and swale (Caño Matanzas site) with the early date obtained by Parsons and Shlemon. Excavation of the Peñon del Río raised fields in conjunction with the deep settlement site adjacent may resolve the question of field age and earliest cultural affiliation.

## Acknowledgements

The principal investigators for the Samborondón Project were William M. Denevan and Donald E. Thompson. Graduate student field directors were Kent Mathewson and Richard G. Whitten. Field assistants were gradute students David Stemper (archaeology) and John M. Treacy (geography). All were at the University of Wisconsin, Madison. Research was supported by the National Science Foundation, the Fulbright-Hays Program, and El Programa de Antropología para El Ecuador, Quito, under the direction of Presley Norton. For substantial assistance and for courtesies extended, we would particularly like to acknowledge the Museo Antropológico del Banco Central, Guayaquil, under the direction of Olaf Holm, and the Excuela Téchnica de Arqueología, Escuela Superior Politécnica del Litoral, Guayaquil, under Jorge Marcos. Portions of this paper were presented at the Conference on Prehistoric Agriculture in the Tropics, Australian National University, Canberra, August 1981.

## Bibliography

ARMILLAS, P. 1971, Gardens on Swamps, Science 174, 653-661.

BATCHELOR, B.E. 1980, Los camellones de Cayambe en la sierra de Ecuador, América Indígena 40, 671-689.

BENDER, M.M., D.A. BAERREIS, R.A. BRYSON and R.L. STEVENTON 1981, University of Wisconsin Radiocarbon Dates XVIII, Radiocarbon 23, 145-161.

BOOMERT, A. 1976, Pre-Columbian raised fields in coastal Surinam, Proceedings of the Sixth International Congress for the Study of Pre-Columbian Cultures of the Lesser Antilles, Guadeloupe, 134-144.

DAMP, J.E., D.M. PEARSALL, and L.T. KAPLAN 1981, Beans for Valdivia, Science 212, 811-812.

DENEVAN, W.M. 1966, The Aboriginal Cultural Geopgraphy of the Llanos de Mojos of Bolivia, Ibero Americana 48, University of California Press, Berkeley, 185pp.

DENEVAN, W.M. 1982, Hydraulic Agriculture in the American Tropics: Forms, Measures and Recent Research", In, Maya Subsistence: Studies in Memory of Dennis E. Puleston, edited by K.V. Flannery, Academic Press, New York, 181-203.

ESTRADA, E. 1957, Ultimas civilizaciones pre-históricas de la cuenca del Río Guayas, Museo Víctor Emilio Estrada, No.2, Guayaquil, 87pp.

KNAPP, G. 1983, Soil, Slope and Water in the Equatorial Andes: A Study of Prehistoric Agricultural Adaptation, Ph.D. Dissertation, University of Wisconsin, Madison, 365pp.

KNAPP, G. and R. RYDER 1983, The use of wetlands in the Prehistoric economy of the northern Ecuadorian highlands, this volume.

LATHRAP, D.W., D. COLLIER and H. CHANDRA 1975, Ancient Ecuador: Culture, Clay and Creativity, 3,000-300 B.C. Field Museum of Natural History, Chicago, 110pp.

LATHRAP, D.W., J.G. MARCOS and J.A. ZEIDLER 1977, Real Alto: an ancient ceremonial centre, Archaeology 30, 2-13.

MARCOS, J.G. (editor) 1982. Projecto arqueológico y etnobotanico Peñon del Río: Informe preliminar y planteamiento de continuación, Escuela Téchnica de Arqueología, Escuela Superior Politecnica del Litoral, Guayaquil, 60pp.

MATHEWSON, K. 1980, Proyecto Camellones: interim progress report, Museo Antropológico del Banco Central del Ecuador, Olaf Holm, Director, Guayaquil, 16pp.

MEGGERS, B.J. 1966, Ecuador, Praeger, New York, 220pp.

NORTON, P. n.d. Early Valdivia middens at Loma Alta, Ecuador, Paper presented at the 37th Annual Meeting of the Society for American Archaeology, Bal Harbor, Florida, 1972.

PARSONS, J.J. and R.J. SHLEMON 1982, Nuevo informe sobre los campos elevados prehistoricos de la Cuenca del Guayas, Ecuador, Miscelanea Antropologica Ecuatoriana 2, 31-37.

PARSONS, J.J. 1969,Ridged fields in the Río Guayas Valley, Ecuador, American Antiquity 34, 76-80.

PARSONS, J.J. and W.A. BOWEN 1966, Ancient ridged fields of the San Jorge River Floodplain, Colombia, Geographical Review 56, 317-343.

PULESTON, D.E. 1977, The art and archaeology of hydraulic agriculture in the Maya Lowlands, In, Social Process in Maya Prehistory: Studies in Memory of Sir Eric Thompson, edited by N. Hammond, Academic Press, London, 449-467.

SANDERS, W.T., J.R. PARSONS and R.S. SANTLEY 1979, The Basin of Mexico: Ecological Processes in the Evolution of a Civilization, Academic Press, New York, 561pp.

SAUER, C.O. 1948, Environment and culture during the last deglaciation, *Proceedings of the American Philosophical Society* *92*, 65-77.

SIEMENS, A.H. and D.E. PULESTON 1972, Ridged fields and associated features in Southern Campeche: new perspectives on the Lowland Maya, *American Antiquity* *37*, 228-239.

STOTHERT, K.E. n.d., Preceramic adaptation and trade in the intermediate area, Paper presented at the 76th Annual Meeting of the American Anthropological Association, Houston, 1977.

TURNER B.L. II and P.D. HARRISON 1981, Prehistoric raised-field agriculture in the Maya Lowlands: Pulltrouser Swamp, Northern Belize, *Science* *213*, 399-405.

VON BUCHWALD, O. 1917, Tolas ecuatorianas, *Physis* (Buenos Aires), 3, 250-260.

WOLF, T. 1892, *Geografía y geología del Ecuador*, Brockhaus, Leipzig, 671pp.

ZEVALLOS, C.M., W.C. GALINAT, D.W. LATHRAP, E.R. LENG, J.G. MARCOS and K.M. KLUMPP 1977, The San Pablo corn kernel and its friends, *Science* *196*, 385-389.

ZUCCHI, A. and W.M. DENEVAN 1979, Campos elevados e historia cultural prehispánica en los llanos occidentales de Venezuela, *Montalban* (Caracas), 9, 565-736.

Chapter 10

# PATTERN ANALYSIS OF PRE-HISPANIC RAISED FIELDS OF LAKE TITICACA, PERU

Thomas J. Lennon

Abstract

This study attempts to understand raised field patterning in the context of hydraulic conditions in the environs of Lake Titicaca, Peru. Previous work on Lake Titicaca raised fields suggested that the irregularity of field dimensions and absence of major canals reflected both the lack of high level of field organisation and the lack of an integrated drainage or irrigation system. The present research identifies raised field characteristics which can be measured and examined statistically within two specific water regimes, the lakeside flood plain and a river flood plain.

Analysis of large-scale aerial photography coupled with intensive pedestrian survey of the study areas provided the data used to describe field characteristics and the associated cultural features. Statistical examination of field characteristics clearly establishes a relationship between variation in field characteristics and distance to lake or river. Additionally, a variety of water control structures, including main canals, feeder canals, improved channels, embankments and habitation mounds were found and are interpreted to have formed an integral part of the raised field complex in both study areas.

Results of this study indicate that the Lake Titicaca raised field patterns and associated features were part of a highly organised, multifunctional water management programme adapted to specific river and lake water regime characteristics.

Resumen

Este estudio intenta comprender la distribución espacial de los campos elevados en el contexto de las condiciones hidráulicas en los alrededores del lago Titicaca, Perú. Trabajos previos sobre los campos elevados del lago Titicaca sugieren que la irregularidad de las dimensiones de los campos y la ausencia de canales mayores reflejaba tanto la falta de un alto nivel de organización de campos como la ausencia de un sistema integrado de drenaje o de riego. Este trabajo identifica características de campo elevado que pueden medirse y examinarse estadísticamente dentro de dos regímenes acuáticos específicos: el llano de inundación del lago y un llano de inundación fluvial.

El análisis de fotografías aéreas a gran escala junto con una intensiva inspección en terreno de las areas de estudio proporcionaron los datos usados para describir las características de los campos y sus rasgos culturales asociados. Un examen estadístico de las características de los campos establece claramente una relación entre la variación en las características de los campos y la distancia al lago o al río. Además, una variedad de estructuras de control de agua, incluyendo canales principales, canales de alimentación, cauces mejorados, terraplenes y lomas se encontraron y se interpretan como formando parte integral del complejo de campo elevado en ambas areas de estudio.

Los resultados de este estudio indican que la distribución espacial de los campos elevados y sus rasgos asociados en el lago Titicaca formaban parte de un programa de manejo de agua multifuncional y altamente organizado, adaptado a características específicas de régimen fluvial y de agua lacustre.

Archaeological interest in studies of prehistoric agricultural practices has been greatly stimulated by the identification of Pre-Columbian agricultural field forms identified as raised fields.* Raised field agriculture includes the construction of mounds, platforms, and ridges usually in association with a variety of cultural features such as canals, dikes, and habitation mounds. One of the largest areas of raised fields in Latin America has been identified in the Lake Titicaca region of Peru and Bolivia. Based upon preliminary air photo studies, the lake plains are estimated to contain 82,056 ha (202,796 acres) of remnant raised fields (Smith _et al_. 1968:355). In addition to the enormous area they encompassed, the Titicaca fields are also notable for the variety and complexity of field forms. The largest concentration of fields is located within Peru and is identified as the Juliaca-Paucaucolla main block containing 356 square kilometres of fields (222 square miles). The present study is located within the Juliaca-Paucaucolla main block (Figure 10:1). Around the lake region, raised field patterns are found mainly on poorly drained flat land at an elevation of between 3,800 and 3,890m and within a 30 km distance from the lake margin (Smith _et al_. 1968:355). Considerable diversity in raised field patterning was noted throughout the lake region.

With regard to raised field function, Smith _et al_. (1968) suggest that they were constructed to reclaim marshland for cultivation or to protect farmland from occasional inundation by rainfall, runoff or from overflowing rivers and lakes, and that there is no real evidence of systematic drainage for the fields.

Pattern classification focused on the field characteristics of length, width, shape, and arrangement, and resulted in the identification of five patterns: open checkerboard, irregular embankment, riverine, linear, ladder, and combed. The author argues for patterning which focuses on water conservation functions, noting that drainage would not be facilitated in a ladder or embankment pattern.

---

*A raised field is defined as "any prepared land involving the transfer and elevation of soil above the natural surface of the earth in order to improve cultivating conditions" (Denevan and Turner 1974:24).

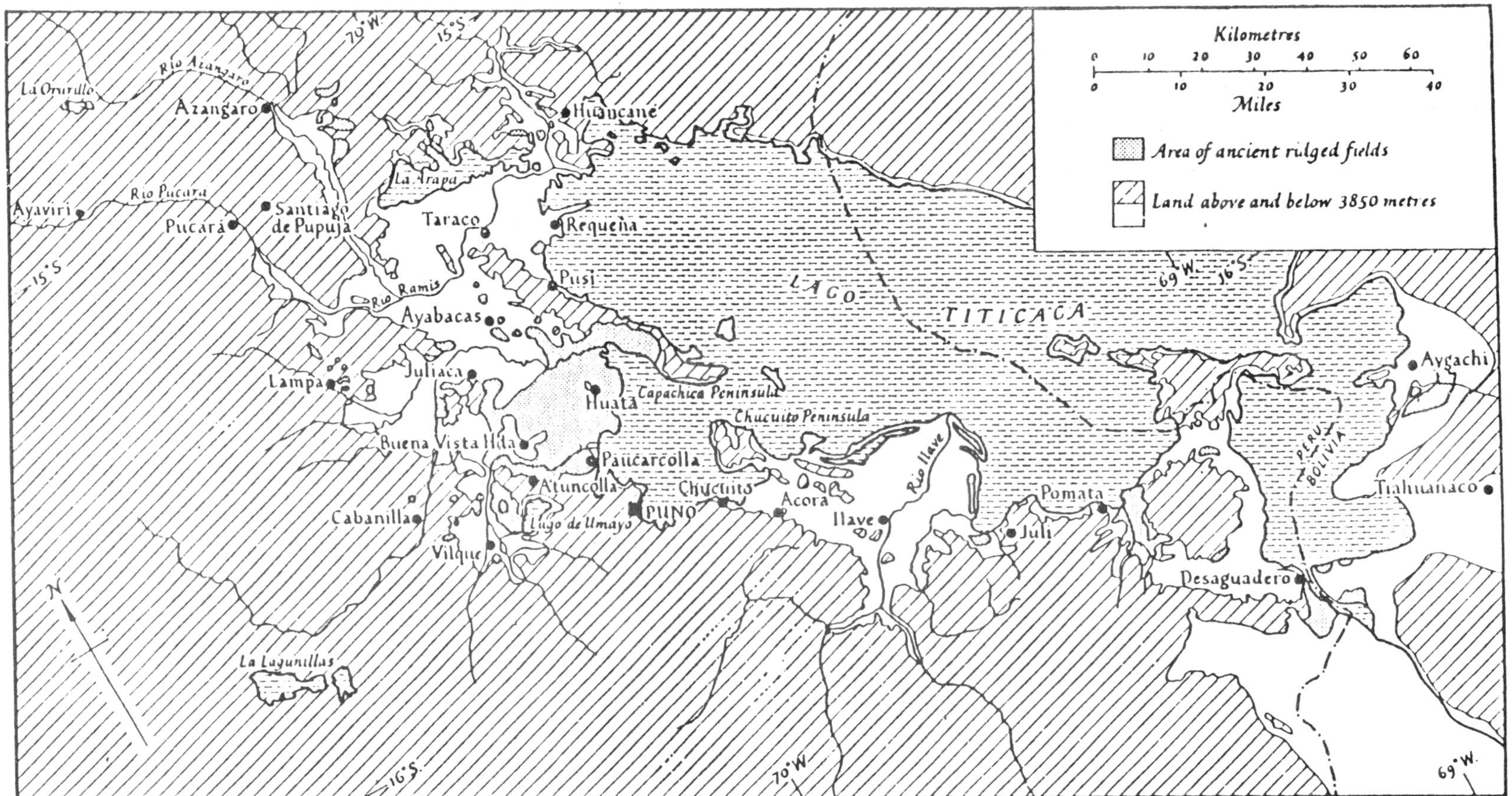

From Smith, Denevan, Hamilton 1968: 354; fig. 1

Figure 10:1 Location of general study area: Jullaca-Paucarcolla main raised field concentration

The argument in support of a water conservation function for field patterns is based on the effect of critical environmental factors: 1) the seasonal and unreliable precipitation pattern in the area, and 2) drought and frost hazards.

The goal of this study is to better understand the raised field patterning in the context of the hydrologic conditions in the environs of Lake Titicaca. The emphasis is on the study of both the agricultural system (raised fields) and the water management system employed in conjunction with field patterning. The present study takes advantage of the excellent air photo coverage of portions of the Titicaca raised field area. The fact that over 400 sq km of field patterns are accessible for study without the masking effect of dense vegetation, as in the lowlands of Mexico, Belize, and Guatemala, makes this a unique opportunity to study specific details of a raised field system within the context of different hydraulic regimes.

An assumption of this study is that the existence of tens of thousands of raised fields along the edge of Lake Titicaca is evidence of an organised Prehispanic response to water resource characteristics specific to the study area. Water resource availability in various hydraulic regimes is seasonal and highly variable from year to year. The study of raised field patterns and their relationship to hydraulic characteristics is, therefore, likely to provide an insight into field form and function and the dynamics of raised field patterning.

Lennon (1982) identified certain parameters of social behaviour and response to water availability (flood-drought) for both lake and river contexts within the study area. On the basis of the described water dynamics and culturally adaptive qualities, several hypotheses can be stated. The hypotheses are based upon the following considerations:

1. Water as a resource is subject to extreme temporal and spatial irregularities of occurrence.

2. Man acts in an organised social and economic fashion in an attempt to exploit resources.

3. Characteristics of water (availability, flow, volume, directionality) within major hydraulic regimes (river, lake) exert a major influence on human use of the resource and other resources (e.g. land) affected by water.

4. Cultural perception of, and adaptive response to, water resource characteristics, can be inferred from the spatial distribution of archaeological data. Archaeological data are defined here to include the following: agricultural field patterns, canals, diversions, embankments, mounds,

channel improvements, reservoirs, and other evidence of Prehispanic landscape alterations occurring within the study area.

These general statements can be reformulated into hypotheses that concern field pattern characteristics in relation to specific hydraulic regimes.

Two general hypotheses can be formulated:

1. If hydraulic regime (river, lake) characteristics exert an influence on raised field patterning, then as the given regime changes there should be changes in the characteristics of raised field patterns.

2. If hydraulic characteristics of a given regime (river, lake) exert an influence on raised field patterning, then as the distance to hydraulic regime (river, lake) changes there should be a corresponding change in the raised field patterning.

These general hypotheses can be reformulated into a series of hypotheses which relate to specific field and feature characteristics. The hypothesis examination format will be based on use of the null ($H_o$) and alternative ($H_a$) format. Rejection of the null hypothesis will provide the basis for identifying and explaining water mangement strategies per hydraulic regime.

The first general hypothesis focuses on interregime adaptation and is reformulated into a series of hypotheses which separately focus on: 1) pattern variation along an axis perpendicular to the lake, and 2) pattern variation along axes perpendicular to the river. Associated cultural features are also identified within each study area.

## Methodology

### The data base

Specific characteristics were selected for physical attributes, spatial orientation, and relation to water of field patterns and associated features. The characteristics chosen for the study area are described in the following manner.

In order to quantify field pattern characteristics, it is first necessary to define a field pattern. A practical approach is to select the smallest identifiable grouping (or "bundle") of individual fields as the unit of measurement. For the purposes of this study, field pattern bundles are labelled sample unit components, and field pattern is used in reference to trends and variation in component characteristics. Descriptors used to describe or measure component characteristics are shape, boundedness, orientation to component axis, and component measurements.

Shape refers to the general component outline such as quadrilateral, ovoid, and trilateral. Boundedness refers to the presence or absence of an embanked border surrounding all or a portion of a component; three levels of boundedness were identified: bounded, unbounded, and semibounded. The orientation of component axes refers to the linear position of the individual fields within a component; reference to: the major hydraulic feature (lake or river), the closest field canal or improved channel, or the orientation to adjacent components. A number of individual measurements were taken in order to identify the dimensions of specific variables or the proximity of a particular component to a hydraulic feature. These measurements are: component long and short axis, length, width and swale, number of fields, and area.

Asssociated features were identified by type and location per hydraulic regime. Associated features are man-made structures or man-altered natural features which represent human adaptation to flood plains. Characteristic flood plain structures and features are embankments, channel improvement, diversion measures, reservoirs, and habitation areas.

Sampling Method

Two study areas were selected within the Juliaca-Paucaucolla raised field block. One study area, labelled the river study area, centres on the flood plain of the Rio Ilpa near the town of Hatuncolla (approximately 15 dm west of lake edge) the other study area, labelled the lake study area centres on the Lake Titicaca flood plain (approximately 1 km north/northwest of the town of Paucarolla). The two study areas lie approximately 0.5 km apart. These areas were selected because of the author's familiarity with the area, the quality of available air photographs, and the integrity of the field patterns observed.

Air photo coverage was made available through the office of the Ministerio de Aeronautica, Direccion General de Aerofotographia, Servicio Aerofotographia Nacional in Lima, Peru. In order to analyse the air photographs, a series of steps was taken which permitted organised and detailed examination of the photography. After scrutinising the available coverage, the appropriate photos were selected for study, enlarged to 18" x 18" size and scaled for both the river and lake study area.

A sytematic sample design was used for both hydraulic regimes. The lake flood plain area was sampled along an axis from the present lake shoreline to a point 6.23 kilometres inland. Each of the seven lake study area sample units measures 24.42 ha. A total of 170.9 ha of lake area was mapped and analysed. The river flood plain was divided into two subareas: 1) the river channel and its immediate environs, and 2) the flood plain lying away from the channel. The river regime was sampled using several axes which crossed the subareas perpendicular to the river channel. Seven sample units located along a three-mile length of the river contain 22.1 ha each, while 13 flood plain area sample units each contains 5.56 ha. A total of 226.98 ha of the river area was mapped and analysed.

The objective of mapping the sample units was to identify, map, and measure all the field variables within each sample unit component. The seven sample units within the lake plain contain 781 components comprising from 90 to 136 components per sample unit. The twenty river sample units contain a total of 344 sample unit components. The number of components per river regime sample unit ranges from seven to 36 components.

Dimensions were measured from the air photograph enlargements (18" x 18") by using a mylar overlay scaled in one-hundreths of an inch. This device permitted accurate measurements within field pattern boundaries and between individual fields. Fine detail measuring was made easier by placing photos on a light-table. The smallest possible measurement unit within the river regime was 2.65m, while within the lake regime the smallest unit of measurement was 1.4m. In this manner, sample unit components were completely outlined on air photos, then measured and recorded, and sample unit detail maps drawn.

All variables were coded onto computer data sheets and processed using the Statistical Package for the Social Sciences (SPSS) language (Nie *et al*. 1975; Klecka *et al*. 1975). Measurements of central tendency and variability (mean, standard deviation, and range) and frequencies of occurrence distribution provide descriptive summaries of sample units. In order to examine the interaction among multiple variables, contingency table analysis, chi-square statistic, and two-way analysis of variance statistics were used.

These statistical approaches allow us to:

1. identify the existence of field pattern variation;

2. summarise field characteristic variation and distribution in relation to distance to water; and

3. test significance by establishing a statistical base for use of the null hypothesis which postulates that, "there is no statistically significant difference between phenomena which occur by pure chance and the statistically evaluated behaviour of the data as they have been observed by the researcher in a survey or an experimental research design" (Leeds 1974:141).

## Interpretations

Statistical examination of the hypotheses has clearly established with a high degree of probability that there are differences in raised field component distributions within the two hydraulic regimes. It has been shown that distance from river or lake is a critical factor in field patterning. The results of this statistical analysis are described elsewhere (Lennon 1982: 93-174, 241, 340).

## Lake Regime Characteristics

Lake levels have both short and long term fluctuation cycles. The fluctuations are a result of the quantity of precipitation received throughout the Titicaca drainage basin. Annual fluctuations have been described as ranging from 50 to 100 cm with an average of 80 cm (Smith *et al*. 1968:363). Multiyear cycles have been described but are based on a very limited record. To date, the extreme levels recorded range from a high of 124 cm in April 1933 to a low of -371 cm in December 1943. This represents a 495 cm vertical difference in lake level in a ten year period.

The effects of both annual and multiyear lake level fluctuation must be taken into account when discussing raised field patterning. Implied in the cycles is the effect of the temporal and spatial distribution of precipitation on the lake edge and the lake flood plain (Lennon 1982:53-73).

Lake level fluxes are the result of the cumulative effects of precipitation throughout the Titicaca drainage basin. Fluctuating lake levels affect varying portions of the lake plain depending on the movement of surface and subsurface lake water. Other portions of the lake plain, those areas far enough away from the effects of the lake level fluctuation, would be more affected by the amount of rainfall, its temporal and spatial distribution, and surface runoff conditions.

A generalised continuum of water availability effects can be described as ranging from short and long cycle drought to short and long cycle flooding for both the lake edge and the lake plain. Long cycle drought and flooding are extreme events which would have devastating effects on the use of raised fields. Long-term drought conditions would severely impact the lake plain and reduce the amount of land available for cultivation down to those watered portions of the lake edge. Long cycle flooding would inundate extensive portions of the lake edge and presumably, as a result of heavy rainfall, waterlogged soil conditions would prevail in the lake plain. The short-term effects of flood and drought would have a greater impact on the lake plain than the immediate lake edge. Seasonal drought or even irregular periods of precipitation can ruin a crop. Short-term floods can be localised and vary in magnitude and duration, but clearly would have the potential to damage both fields and crop.

Given the variability of available water resources, there would clearly be a need to facilitate water conservation during drought periods and drainage during periods of excess water. These conditions provide a general scenario within which the lake area patterns can be summarised.

## Lake Summary

Three basic lake area water resource zones can be defined: 1) lake edge (fluctuating), 2) lake plain surface runoff area, and 3) area of combined effect. The lake edge zone is the perimeter of the lake edge and is subject to the range of lake level fluctuations. During extended dry periods the lake edge zone would retreat to lower levels and expose areas which usually lie under lake water. Large areas of submerged raised fields noted in air photographs taken in April 1977 were probably constructed during such a dry period. Although subject to the effects of fluctuating lake levels, this zone would continue to be a most important resource area due to the combined availability of year round water, fish, birds, and aquatic plants. Given the widespread impact of both short and long term drought on extensive areas of the lake plain, the lake edge serves a critical need to secure a supply of water which would be available season after season for cultivation and other subsistence practices. The surface runoff zone is that lake plain area which exists beyond the effect of the fluctuating lake zone and which relies upon the collection and distribution of surface water runoff for cultivation. This area is generally better drained than lake edge land, but also more susceptible to drought conditions. The combined effects zone is the low-lying area inland from the lake which can be subject to lake edge and surface runoff conditions.

The study of raised field patterns within the lake study area indicates a fairly consistent trend in the variation of field dimensions. The trend indicates that larger dimensions are generally found away from the lake while smaller dimensions are found closer to the lake. Within the lake edge and surface runoff zones, another pattern can be identified. There is a noticeable subdivision of larger and smaller mean dimensions. For instance, within the high water availability lake edge area, there is an emphasis towards proportionally larger means to be grouped closer to lakeside, and the smaller means to be found further from the lake. The same is true for the low water availability lake plain. This range of variation within the lake study area may be indicative of planning within common flood stage parameters or be representative of minor or topographic variation within the sample unit (e.g. low-lying areas).

Many ancillary structures have been identified with the lake sample area which facilitated the use of raised field patterns within the continuum of water resource availability. Numerous flood reducing structures have been identified in the flood plain. Improved channels, large canals, and feeder canals have been identified which indicate attention to water distribution through and around the field components as well as the need for direct removal of excess amounts of surface water. Two areas, the lake edge and a habitation mound concentration located near the sample units, show a heavy emphasis on the use of embankments. These locations are the most highly landscaped portions of the study area. Numerous circular and semicircular embanked empoundments found in low lying areas may have served as water conservation structures. Many large habitation mounds are surrounded by extensive diking systems, which in many cases are multitiered arrangements.

In summary, the data have shown that the lake study area contains evidence of water distribution and conservation measures, in addition to drainage measures, which indicate an overall planning for a range of water resource availability effects.

## River Regime Characteristics

Very little data exists for river level fluctuation in the Lake Titicaca region. Certain characteristics can be described based on the limited data available. The rainy season accounts for most of the annual river volume, while dry season river volume is attributable only to snow melt from the more remote areas of the drainage basin. Fluctuation in river volume for both short- and long-term cycles can be related to the amount of precipitaiton, and its temporal and spatial distribution.

Variation in rainfall distribution can occur in all or a portion of the drainage basin, affecting both river levels and river plain surface runoff. The magnitude of river area flooding is dependent upon soil moisture, duration of and intensity of rainfall, and extent of area affected. Based on these conditions, a continuum of flood and drought effects ranging from short-term (irregular rainfall or single even floods), to long-term (multiyear drought or seasonal floods) can be described. Long-term flooding would severely affect raised fields throughout the river area. Short-term flooding, depending on the magnitude, could equally affect both the river flood plain and the river channel area. The long-term affect of drought would be devastating throughout the river area. River channel volumes during the wet season could be reduced to nothing, if the drought had an affect throughout the area drained by the river. Short-term drought effects would have a greater impact on the river flood plain. Irregular precipitation could severely affect plant growth early in the growing season.

Given the nature of water resource variation, there would clearly be a need to facilitate water conservation during drought periods and drainage during periods of excess water. These conditions provide a general scenario within which the river area patterns can be discussed.

River Summary

Three basic river area water resource zones are defined: 1) river channel zone, 2) surface runoff, and 3) combined effects zone. The river channel area contains the most available water, the best soil in the area, and also has good drainage conditions. The surface runoff zone is that area which is more affected by runoff conditions than by channel flooding. This zone also has generally good drainage due to the relatively higher elevation. A combined effect zone is that area of the flood plain which is subject to the accumulation of surface water runoff and flood waters. Inundation, caused by short or long-term flooding, would be the major problem here due to the low-lying nature of the area. In periods of short or long-term drought, this area would be a natural catchment for available surface runoff, and serve as a critical buffer zone for maintaining agricultural production during periods of diminished water availability.

The study of raised field patterning in the river regime has indicated a very strong dichotomy in field patterning. Component dimensions decrease in areas of poorer drainage and increase in areas of better drainage. By constructing smaller patterns in low-lying areas, more drainage area is created in and around field components. Adding to more emphasis on drainage is the closer proximity to field canals located in the poorer-drained areas. Low-lying areas would be expected to be the last field area to be affected by drought conditions and the first area to be impacted by flooding conditions. Relatively higher and better drained land at the river edge and near the hill slopes would be impacted first by drought conditions and would be the last area to be impacted by large volumes of surface runoff.

Complementing this multipurpose arrangement is the diversity of water control structures evident throughout the study area. There is strong evidence for a surface runoff collection system which could gather surface runoff in areas close to hill slopes, into major canals and divert it away from the lower-lying areas.

The land at the break of hill slopes has a standard arrangement of large raised fields arranged perpendicularly to the base of the hill. These fields are, in turn, generally surrounded by a large canal which surrounds the hill slope base and then proceeds to move through the flood plain and into the river channel. This pattern has been identified on both sides of the river. In other relatively elevated areas not near hill slopes, surface runoff would move through the maze of components apparently at a slower rate than on lower-lying fields.

It is clear that specific portions of the river flood plain are designed more for water drainage than for conservation functions. This planning effort is focused in the area immediately adjacent to the river channel. There are large canals through the flood plain paralleling the river channel apparently to facilitate movement of large volumes of water. At certain points these canals feed back into the river channels. Planning for conservation of water resources for these areas has also been discussed. The existence of one very long diversion channel is an indication of a major effort to divert the river flow through an extensive area of the flood plain.

There is no evidence of special structures constructed to protect the river bank area from erosion although at critical stress points where erosion would be heaviest, field components are generally oriented perpendicularly to the channel. Habitation areas were located at higher points along the river edge. Some smaller mounds exist in the flood plain, but there is no evidence of intensive occupation of the river flood plain.

In summary, the river flood plain is extensively landscaped in a manner that indicates a response to the range of effects of water availability within different areas in the flood plain.

Comparison of Lake and River Study Areas

The existence of a variety of field design characteristics and associated structures within the two study areas tends to support the idea that water resource management was not planned towards strictly conservation, or drainage, or irrigation measures. It appears evident that both lake and river areas were developed with a concern for a range of water availability effects; e.g. short to long-term drought or flood conditions, but with specific planning relevant to the behaviour of each particular hydraulic regime.

Both areas have shown purposeful planning which reflects an understanding of the characteristics of water flow as a major element in the control of water resources. Within each study area the relative range of drainage frequencies can be identified and used in understanding the physical attributes of raised field patterning. Each study area has been shown to represent an attempt to take full advantage of natural features and resource characteristics in order to plan for the extreme variation in water resource availability. In order to do this, field form and associated features were designed to operate within a functional continuum which has been shown to range from water conservation to water drainage. Such planning indicates not only an understanding of water resource characteristics and variability, but represents a decision to design field patterns in a manner that would ensure the use of a least some portion of the lake and river areas for cultivation regardless of the water resource variation.

A comparison of the pattern characteristics of each regime indicates the extent of planning specific to each hydraulic regime. In comparing the distribution of field shapes there is an extremely close parallel in the representation of the two dominant shapes throughout the entire sample population. For the river regime, the distribution was 75.2% quadrilateral components and 23.9% ovoid, while for the lake regime, 75.4% were quadrilateral components and 21.8% were ovoid. These figures may represent certain land use or land ownership qualities. In reviewing the use of bounded fields, the almost total lack of representation in the river area contrasted to the well-developed embankments in the lake area and this is of interest. The absence of embankments may be due to the seasonal availability of water in the river flood plain as opposed to the long-term availability of water at the lake edges with more labour invested in areas that have more available water. Table 10:1 compares the mean dimensions of certain characteristics of both the regimes. It is intriguing that the river area contains larger field patterns overall and the lake regime is consistently smaller in dimensions with the exception of the land to water proportion.

The clear difference in field design proportions between the two regimes further supports the argument for comprehensive water management planning in the study area. The consistent difference in dimensions may serve as an indicator of the overall water availability characteristics per regime. The river regime, with its emphasis on larger dimensions, would indicate an overall planning approach to the river regime as a low water availability regime relative to the high availability emphasis of the lake plain and lake edge, as indicated by the smaller dimensions. In terms of understanding long range adaptive strategies for water resources, we can describe the river regime as a resource planning area that is characterised by less water availability overall than the lake regime. The river regime can be considered as a less stable water resource base subject to more impact from drought and flooding than the lake area. Emphasis has been placed on critical considerations such as temporal and spatial irregularities of resource occurrence, resource characteristics within different hydraulic regimes, and adaptive responses to resource conditions.

The operational principle underlying the management system can be explained in terms of adaptation to both large and small scale water availability values. Field patterns reflect an adaptation to surface runoff volume variation found to exist naturally within the study areas. The compactness of surface water drainage networks, as

| Category | Regime | |
|---|---|---|
| | Lake | River |
| Long Axis | 53.02m | 84.07m |
| Short Axis | 30.66m | 46.27m |
| Field Length | 32.22m | 51.07m |
| Field Width | 3.48m | 4.56m |
| Field Swale | 2.20m | 3.28m |
| Area | 1038.96 sq.m | 2339.97 sq.m |
| Proportion: Land-Water | 1.74 | 1.53 |
| Distance to Canal | 129.18m | 139.40m |
| No. of Individual Fields per Land Unit | 13.25 per acre/ 32.74 per ha | 3.99 per acre/ 9.86 per ha |
| Average Number of Fields per Component | 7.59 | 8.27 |

Table 10:1 Comparison of Hydraulic Regime: Field Pattern Characteristics

represented by field and component dimension variation, is shown to be a key factor in the overall raised field patterning.

## Summary and Conclusions

### Assessment of Smith, Denevan and Hamilton (1968)

The results of this study are useful in assessing and clarifying the preliminary observations of the Lake Titicaca raised fields proposed by Smith *et al*. The authors stressed the need to understand the diversity of field patterning with particular emphasis on raised field function and the antiquity of raised field cultivation in the Lake Titicaca region. Their initial attempt to describe field pattern characteristics in terms of potential water conservation or drainage functions was instrumental in the development of the research design of the present study. With the combined advantage of more appropriate scales of aerial photography, extensive field survey, and detailed data analysis, it is possible to address the preliminary field observations of Smith *et al*.

It was originally suggested by Smith *et al*. that the field patterns do not represent an organized attempt at drainage, water conservation, or irrigation. Most field patterns were thought to have hindered systematic drainage, not facilitated drainage (1968:261). It was argued, therefore, that raised field patterning represents more of a loosely structured attempt at water conservation due to the seasonal and unreliable precipitation in the areas and drought and frost hazards. Based on their preliminary field survey near the present river study area, spatial variation in field size and field numbers served as an indication that reclamation could not have been "highly organized or stringently planned" (Smith *et al*. 1968:357). The absence of well-defined major canals suggested to the authors an emphasis on increasing the amount of cultivatable land, as opposed to an emphasis on an integrated drainage or irrigation system.

The present study has shown that the variation in field patterning exists and can be explained in terms of surface drainage characteristics. What was originally viewed as an irregularity in dimensions, has been shown to represent a purposeful adaptation to an apparently perceived hazard. The variation in size, described by Smith *et al*., was based on a unit measurement taken of average field width and trough. The authors note that this measurement of ridge and trough or *wave-length* varied greatly in the areas they investigated (1968:357). For instance, between the towns of Vilque and Hatuncolla, the average wave-length was 6.5m; a few miles to the north the wave-length increased to 15m. Near the Hacienda Machacmarca, the average was 10 to 12m, while on the southern shore of the lake, the average was 10.5 to 14m.

The present study has shown that by the use of the wave-length measurement originally discussed in Smith *et al*. (1968), in conjunction with other measurements, that dimensions tend to vary with the general surface runoff characteristicts of an area. What Smith *et al*. originally construed as irregular variations throughout the plains, can now be explained in terms of apparently systematic attempts to organize surface runoff in the manner which paralleled the natural surface drainage density and took advantage of the runoff for agricultural production, thus making more land cultivatable.

The results of the present study have found evidence for major canals. Major canals have been identified throughout the river and lake study areas. Major canals would identify a main link in a water control or distribution system as opposed to a lesser canal form, such as a feeder canal common throughout the study area. The canals and the large diversion structure discussed in the river study would have served to move excess water from one area to another for either drainage or water distribution functions. The fact that major canals have been identified does not, however, imply that there was any less of an emphasis on the amount of cultivable land. The results of this study show tht the objective of field patterning was to develop cultivable land in as many areas as possible. The key to success, however, was designing field patterns and associated features which accommodated the natural behaviour and availability of surface runoff and lake water.

The results of the present study also indicate that the embankment earthworks are an intensive, small scale solution to water resource fluctuations and water control needs in the lake area. Smith *et al's*. comment that long continuous earthworks would provide adequate protection for large areas in periods of high water is considered an impractical solution to the complex problem of water resource variability.

Given the extent of field pattern distribution, Smith *et al*. suggests that the present lake level is similar to past lake levels. This is undoubtedly true but the range of fluctuation of lake levels is a critical variable. Areas of raised fields become visible as the lake edge recedes during the dry season indicating the development of lake edge fields beyond the present level. Raised fields are also found in the bottom of small lakes and in low-lying areas on the tops of hills bordering the river study area.

The present study indicates that the need for protection was addressed in the design of embankments, canals, improved channels, and diked habitation mounds; that is, in a manner which worked with water characteristics and avoided designing against natural forces. Protection was designed on a smaller scale because that scale was most effective in directing and controlling water given the Prehispanic perception of the hazard. This study has described the existence of major and minor structures associated with the field patterns which are the whole water management system. The interaction of these components has been shown to serve both conservation and drainage functions in an integrative manner. That is, there is ample evidence for defining a water management system which cannot be labelled as either a water conservation drainage or an irrigation system, but must be understood in terms of being designed for multiple response to variation in water resource characteristics. The system was apparently planned to provide a continuum of conservation and drainage function within each hydraulic regime.

In summary, the Lake Titicaca raised fields have been shown to be an elaborate Prehispanic water management system characterised by extensive area, complexity, and a high degree of organisation. The creation and maintenance of such a system in an area only marginally suited for agriculture has tremendous implications for the study of Prehispanic sociopolitical development in the Titicaca basin.

## Acknowledgements

Support for this research was provided by the National Science Foundation (Dissertation Improvement Grant, BNS 76-19861); the Fulbright Commission, Lima; Instituto Nacional de Cultura; and Museo Nacional de Antropologia y Arqueologia.

## Bibliography

DENEVAN, W.M. and B.L. TURNER II 1974, Forms, functions and associations of raised fields in the Old World tropics, Journal of Tropical Geography 39, 24-33.

KLECKA, W.R., N. NIE and C. HADLAI HULL 1975, Statistical Package of the Social Sciences Primer, McGraw-Hill Book Company, New York,

LEEDS, D. 1974, Practical Research: Planning and Design, MacMillan Publishing Co., Inc., New York.

LENNON, T.J. 1982, Raised Fields of Lake Titicaca, Peru: A Pre-Hispanic Water Management System, Ph.D. Dissertation, University of Colorado, Boulder, 340pp.

NIE, N.H., H. BENT, and C. HADLAI HULL 1970, Statistical Package for the Social Sciences, McGraw-Hill Book Company, New York.

SMITH, C.T., W.M. DENEVAN and P. HAMILTON 1968, Ancient ridged fields in the region of Lake Titicaca, Geographical Journal 134, 353-367.

Chapter 11

# ASPECTS OF THE ORIGIN, MORPHOLOGY AND FUNCTION OF RIDGED FIELDS IN THE QUITO ALTIPLANO, ECUADOR

Gregory Knapp and Roy Ryder

## Abstract

This paper reports the discovery of ridged fields in the Ecuadorean highlands which are of special interest due to their location in the Quito altiplano. The fields are unique because stratigraphical evidence suggest that the abandonment was related to a fall of volcanic ash.

The fields have been largely destroyed but they may originally have covered 18 square kilometres of flat land. No carbon samples or ceramics were recovered but the features appear to be related to the nearby late prehistoric Chillogallo settlements.

The raised fields occur as parallel ridges, sometimes arranged in a checkerboard pattern. The wavelength of the ridged fields is 3.5m with a height of over 0.8m. It is demonstrated that an important function of the fields may have been microclimatic management to combat seasonal moisture deficits and frequent frosts. Also, the exceptional height of the ridges may be partially explained by the desire to renew fertility of the planting surface with subsoil and muck obtained from the ditches.

Findings such as this in Colombia, Ecuador and Peru point increasingly to the importance of poorly drained flats as a key niche in prehistoric Anden economies. The Quito raised fields suggest that the choice of this site by the Inca for their northern capital may have been conditioned by ecological as well as military factors.

## Resumen

Este trabajo informa sobre el descubrimiento de campos lomados en las tierras altas del Ecuador que tienen un interés especial debido a su ubicación en el altiplano de Quito. Los campos son únicos porque existe evidencia estratigráfica que indica que su abandono esta relacionado con una caída de ceniza volcánica.

Los campos han sido en gran parte destruídos pero originalmente podrían haber cubierto 18 $Km^2$ de tierra plana. No se recuperaron ni muestras de carbono ni de cerámica pero los rasgos parecen estar relacionados al caserío cercano de fines de la prehistoria de Chillogallo.

Los campos elevados ocurren como lomas paralelas, a veces presentando una distribución en forma de tablero. La longitud de onda de los campos lomados es de 3,5 m con una altura sobre 0,8 m. Se demuestra que una función importante de los campos puede haber sido el manejo microclimático para combatir deficiencias estacionales de humedad y heladas frecuentes. Además, la altura excepcional de las lomas podría explicarse parcialmente por el deseo de renovar la fertilidad de la superficie con subsuelo y estiércol obtenidos de las acequias.

Descubrimientos como éste en Colombia, Ecuador y Perú indican cada vez más la importancia de tierras planas de mal drenaje como una posición clave en las economías andinas prehistóricas. Los campos elevados de Quito sugieren que la elección de este sitio por los incas para su capital del norte podría deberse tanto a factores ecológicos como a factores militares.

## Introduction

Abandoned agricultural ridges (raised fields) have recently been discovered in a number of Ecuadorian highland wetlands. Such fields were first reported in the Cayambe Plain by Ryder (1970) and have since been located in other basins to the north of Quito (Athens 1980; Knapp 1981; Knapp and Denevan 1983; Gondard and Lopez 1981). This paper reports the discovery of a new site which has the southernmost highland raised fields yet to be verified in Ecuador. The site is of special interest due to its location in the Quito altiplano, site of the Inca northern capital, and because of stratigraphical evidence which suggests abandonment related to a fall of volcanic ash.

## The Location and Extent of the Raised Fields

The site is located in an extensive plain south of Quito (Figure 11:1), close to Chillogallo and Barrio Santa Rita at an elevation of 2855m. Approximately 8 hectares of pasture with raised fields can be observed on available air photographs (Figure 11:2: Instituto Geografico Militar "Proyecto Ciudad de Quito" photograph 112, Roll AF-1, Scale 1:6100, November 22, 1968); these features undoubtedly occupied a greater area than this but they have been largely destroyed by ash falls, urban sprawl and pasture improvement. The total area of the plain south of Quito is about 18 square kilometres; since raised fields are not associated with slopes of over 5 percent, this can be considered the maximum possible extent of the original raised field complex.

The "wavelength", or distance from ridge centre to ridge centre, is fairly consistently 3.5m. Excavations have indicated that ditch width was originally equal to ridge width; also the original distance from ditch bottom to ridge top ("amplitude") was found to be over 0.8m. Ridge length varied greatly but typically was between 25 and 75m. If only the ridge tops were cultivated then the width of ditches and ridges implies that the original cultivated surface was about half of the total area of the plain.

## The Cultural and Historical Context of the Raised Fields

No ceramic or other cultural material was recovered from our excavations of raised fields. Unfortunately, it was also impossible to find carbon samples. However, the site is located close to the late prehistoric Chillogallo settlement excavated by Echeverria (1976). The latter was a longitudinal settlement of family nuclei located along

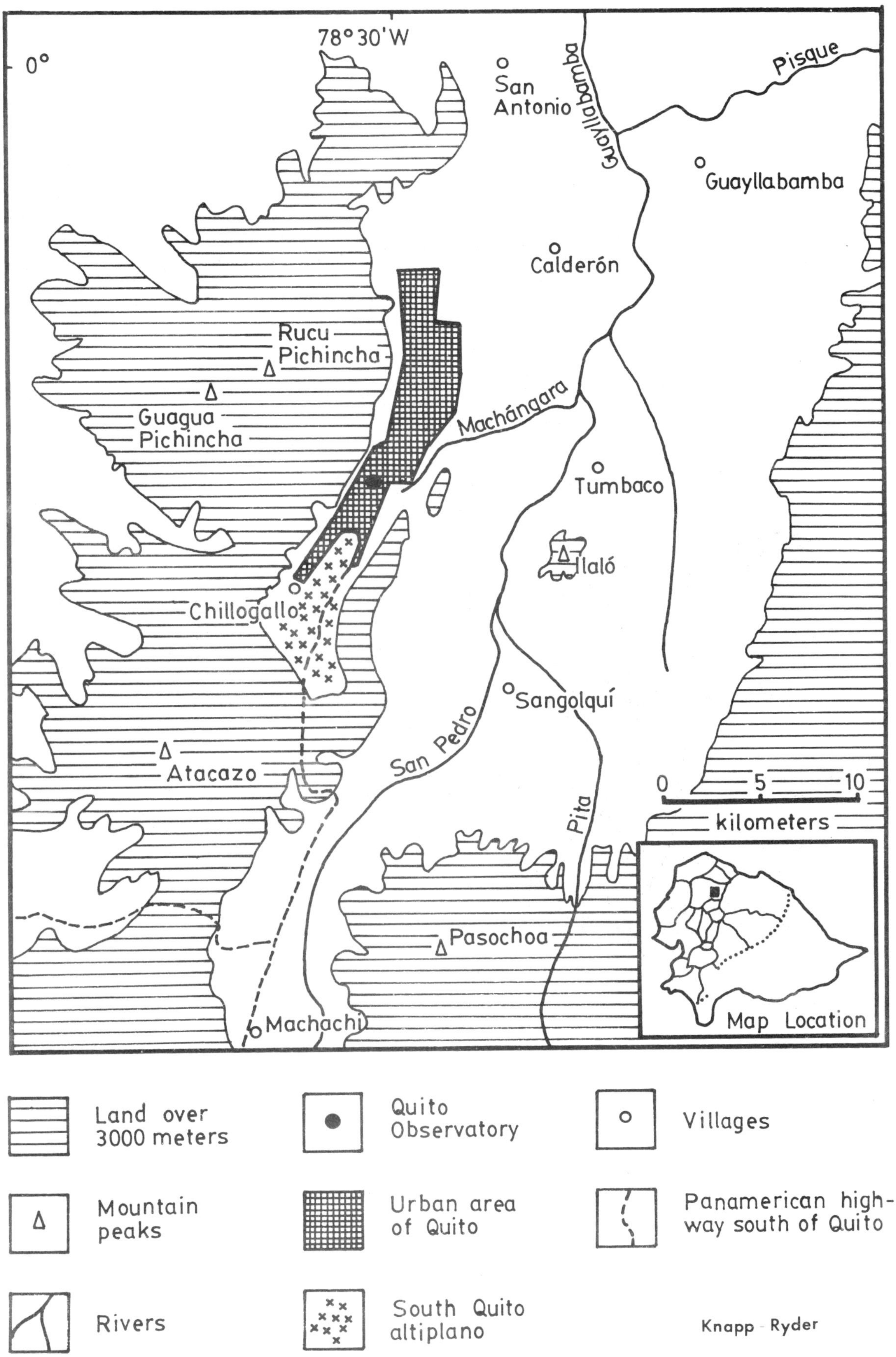


Figure 11:1 Location of the South Quito Altiplano

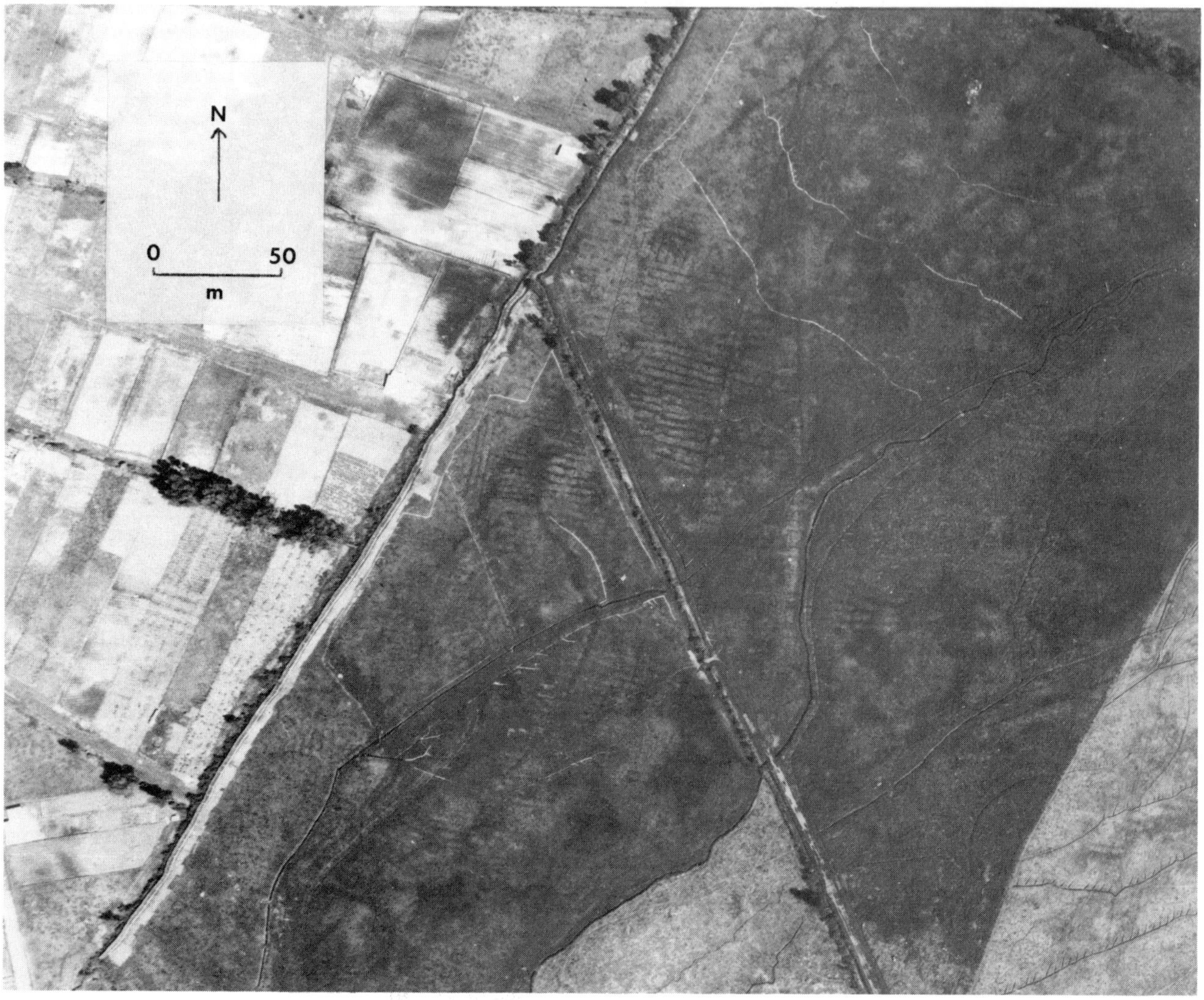


Figure 11:2a Air photograph of flats east of Barrio Santa Rita

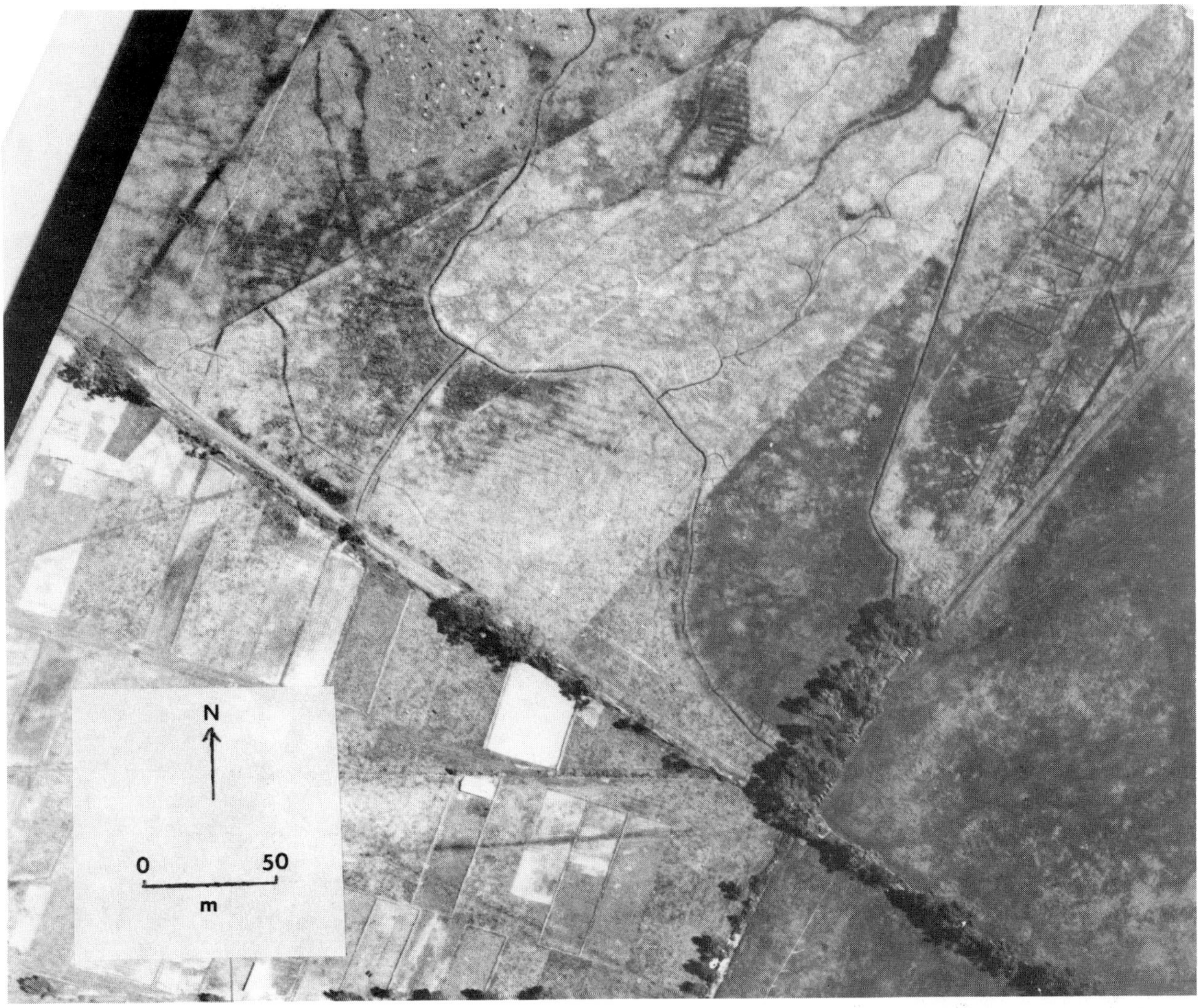


Figure 11:2b Air photograph of flats north of Barrio Santa Rita

<u>Figure</u> <u>11:2c</u> Air photograph of ridged fields in Barrio Santa Rita flats (enlargement of part of Figure 11:2a). Source of photographs: IGM Proyecto Ciudad de Quito, November 22, 1968, Roll AF-1, Photograph 114

the edge of the flats. Abundant burials are evidence of a dense population and cultural remains indicate an affinity with more northerly cultures such as Cayambe, known to be associated with prehistoric raised fields (Ryder 1970). In one case, near San Pablo, northern raised fields are architecturally integrated into a structure which was probably developed in the Late Period, A.D. 1250-1515 (Knapp and Denevan 1983; Athens 1980); thus, a similar date is plausible for the Chillogallo site and associated raised fields.

Immediate post-Conquest accounts describe the flats south of Quito as being marshy and brushy, serving as pasture for Spanish-owned cattle. The indigenous population is described as living on the slopes, farming maize and potatoes in ridges constructed with a foot plough; the ridges were separated from one another by little more than 30 cm (Rodriguez de Aguayo 1965: 202, 204; Anonymous 1965: 212, 223, 227). The scientist-traveller Francisco Jose de Caldas described the same flats as marshy and flooded in the early 19th century (Jose de Caldas 1942: 36). A resident of Barrio Santa Rita who was interviewed during our excavations declared that the area of surviving raised fields was marshy until drained 30 or 40 years ago.

In other areas, it would seem that raised fields were abandoned during the Inca and Spanish invasions due to depopulation and movement away from the strategically vulnerable flats (see, for example, Salomon 1980: 90-95). On the other hand, at the Santa Rita site, ditches contain a lower fill of white volcanic ash (see below) which suggests field abandonment after a still undated volcanic eruption.

The form and size of the raised fields do not correspond to known modern or colonial agricultural technology. For example, when potatoes or maize are grown in modern ridges (<u>huachucuna</u>, plural of <u>huachu</u>, Quichau), the wavelength of the ridges is never over 1.2m, and the depth of the ditch between ridges is never more than 0.5m. Local farmers deny that the Santa Rita raised fields have been farmed in memory and doubt that they were ever farmed in historical times.

Although most of the surviving raised fields have an orientation which is parallel to the gradient, some sets exist perpendicular to the slope and adopt a crude checkerboard pattern (Figure 11:2) which is typical of other presumably prehistoric complexes of raised fields decribed elsewhere in South America (Denevan 1970). The reason for this pattern is unknown; it is an indication, however, that water conservation was more important than drainage to the ridge builders.

## The Physical Context of the Raised Fields

The Quito Observatory (2818m), a meteorological station with long records (1891-1979), is located only 8.5 km N/NE of the Santa Rita raised field site and differs in altitude from the latter by only 37m. Mean annual precipitation at this station is 1222 mm with a standard deviation of 207 mm; annual evaporation varies from under 600 mm in wet years to over 1000 mm in dry years. These figures might suggest that drought risk is negligible, but they do not take into account seasonal variation. In fact, as illustrated in Figure 11:3A, soil moisture deficits develop most years in the dry season (June-September). Agricultural rhythms on the slopes are related to this dry season; maize planting begins in October, and potato double-cropping is impossible without irrigation.

Another problem may develop. In Figure 11:3A, one can observe the lower precipitation in the December-February period, and Figure 11:3B demonstrates that this so-called "little summer" can become severe in some years, leading to soil moisture deficits in the middle of the growing season. This hazard may have been greater in late prehistoric times (Knapp 1980). Thus, it is clear that, even with the high precipitation on the Quito altiplano, there are definite advantages to be obtained from irrigation, and one can appreciate the value of edaphically moist niches such as the raised fields on the poorly drained flats.

The mean monthly temperature of the Quito flats does not vary much throughout the year due to the equatorial location. The mean annual temperature of 13.1°C permits cultivation of maize and potatoes, albeit with long growing seasons; maize requires 10 months and potatoes 4-6 months.

Frost represents a considerable hazard. One measure of this risk is the "median growing period", i.e. the median number of months after sowing that surface minima remain above some defined threshold temperature to prevent serious damage. Table 11:1 shows the results of such calculations for the Quito Observatory for various months of sowing and different definitions of dangerous minimum surface temperatures. It should be emphasized that these calculations reflect conditions on a grassy flat with scattered trees and buildings; microclimatic management could raise surface minima.

Andean radiation frosts typically involve a cold layer only a few decimetres thick; risk to plants is therefore greatest when they are sprouting because they are small and completely within the layer of cold air. Tubers can resist frost better than seed crops, but only before the plant emerges from the soil and during the phase of ripening (Frere _et al_. 1975). Thus, a glance at the table indicates that only January is a normally safe month for

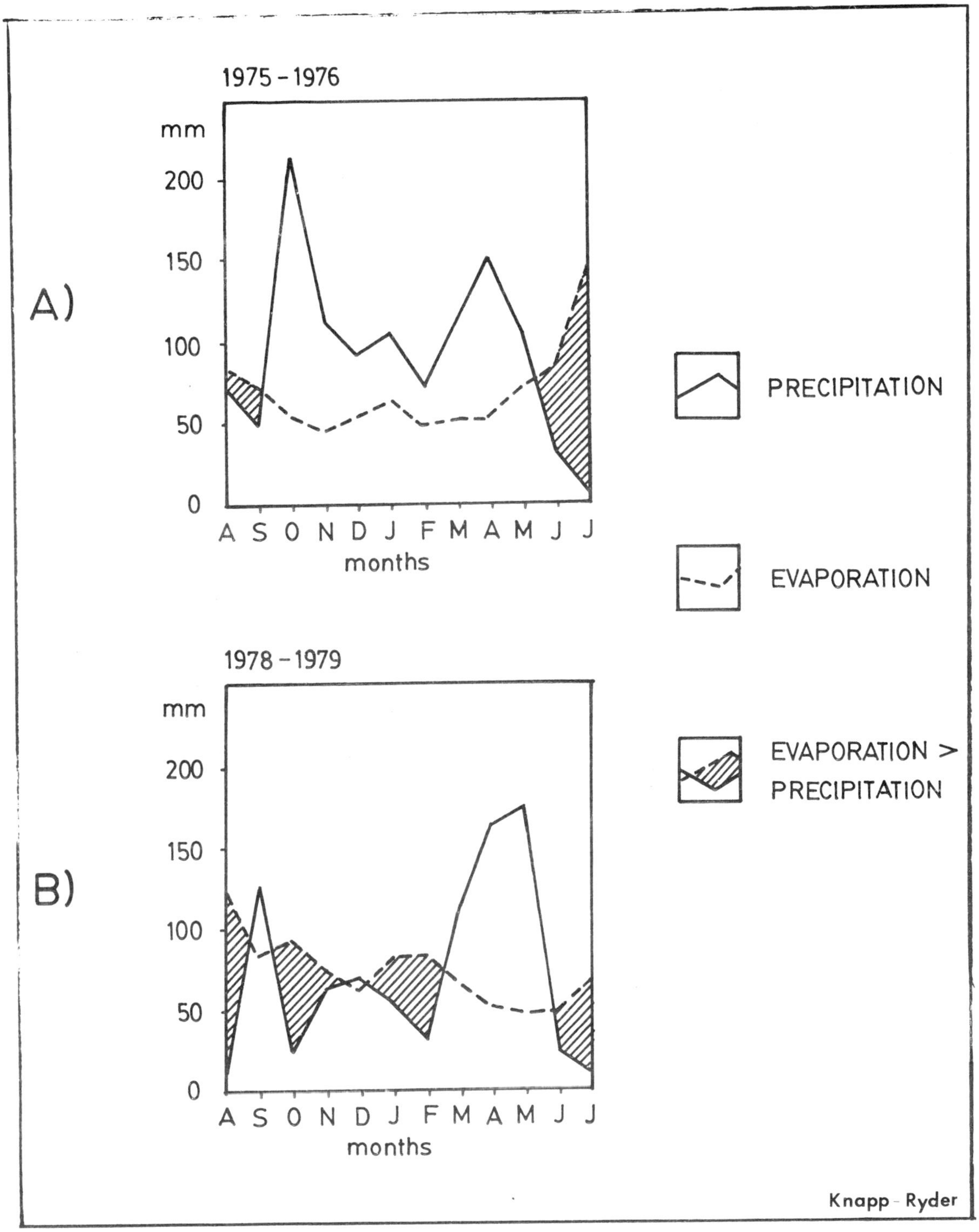


Figure 11:3 Water balance at the Quito Observatory, 1975-1976 and 1978-1979

Table 11:1 Median Lentgh of Growing Period (in months), Quito Observatory, 1937-1966

| Month of | Definition of "Dangerously Low Temperature" | | | | |
|---|---|---|---|---|---|
| | $0^{\circ}$ | $-1^{\circ}$ | $-1.5^{\circ}$ | $-2^{\circ}$ | $-3^{\circ}$ |
| October | 0.0 | 2.0 | 5.5 | 8.5 | 12+ |
| January | 3.0 | 5.0 | 5.0 | 8.5 | 12+ |
| July | 0.0 | 1.0 | 2.0 | 3.0 | 12+ |

Source: Manuscript records of minimum surface temperatures, Quito Observatory

planting in the flats without microclimate management. Planting in October or July should be accompanied by measures to raise nightly minima by 1° to 2°C. Such measures may include ridging, mulching, and deep ditching to create interpenetrating fingers of standing water.

As part of our researches, the senior author created an experimental raised field for frost risk experiments. Three minimum-reading upright thermometers were used; one was placed on a uniform 4 percent slope; another was placed on an adjacent flat. A third thermometer was placed on the bank of a long drainage ditch in the flats. At the latter site, a tank and a connecting ditch were dug to mimic prehistoric raised-field cross-sectional characteristics. Minimum nightly temperatures were read for one week, August 14 to August 20, 1981. Four months later the experiment was repeated, with the same locations as before but with rotated thermometers. In the interval, the tank and supplying ditch had silted up; readings were taken on the bankside during two nights prior to re-excavating the tank. Then the tank and its supplying ditch were re-excavated and six more thermometer readings were obtained (December 23-27). All thermometers were calibrated against a local meteorological station.

During the two nights a thermometer was placed by the field ditch without the construction of a tank to mimic raised field cross-section, the minima of the ditch-side thermometer were within 0.2°C of the control thermometer in the flats. One ditch alone is not sufficient to improve frost characteristics.

The construction of a complete raised-field cross section resulted in clear improvements in minimum temperature (as compared with adjacent flats) in both August and December. Our findings suggested that raised field surfaces were 1.3°C warmer than adjacent flats on nights when the latter experienced frost. This temperature improvement seems to be related more to the development of tubes of warm air over ditches with standing water than to processes of cold air drainage which might be easily disrupted by wind. For the four days when slope minima were the same as or higher than the flat minima, suggesting that conditions were not good for undisturbed cold air drainage, ridge temperatures were still an average of 0.9°C higher than the adjacent flat.

We can therefore conclude that raised fields may enable double-cropping where otherwise crop losses would be excessive, and that the number of growing seasons with frost may be reduced dramatically. Moreover, since the benefit of frost control required deep ditches on both sides (not just one side) of the planting surface, relatively narrow raised fields would have been favoured in areas subject to frost risk. We have here a probable explanation for the relative narrowness of the highland Ecuadorian raised fields as compared with raised fields in the lowland tropics (Knapp 1983).

? B.S.

The soils of the south Quito flats have been classified as typical fertile volcanic soils (Typic Eutrandepts); they are deep, brown-black silt loams, sand loams and clay loams with no limitation for agriculture except poor drainage (PRONAREG-ORSTOM 1980 and authors' observations). Today deep ditches maintain the water table below 1.5m. In a natural state mineral nutrients are abundant and the soils give excellent yields (Frei 1957: 26), although fertilization is necessary for continued cropping. Small farmers fertilize with guinea pig, pig, cattle, burro and sheep manure as well as ash, green manure (remains of weeds, corn stalks and eucalyptus leaves), night soil, and rarely, chemical fertilizer. Considering the absence of manure from large domesticated animals in prehistoric times the exceptionally large amplitude of the abandoned raised fields (compared with contemporary ridges or huachucuna) may be explained in part by the desire to renew fertility of the planting surface with silt and muck accumulated in the ditches (see below).

## Excavation of Santa Rita Raised Fields

Preliminary probes with a soil auger indicated that the ditches between raised fields had become filled with 50 to 90 cm of grey volcanic ash. A trench was excavated perpendicularly across one of the ditches (Figure 11:4) revealing in greater detail the nature of this fill (see also Appendix for soil profile description). The ditch had received volcanic ash from at least two eruptions (horizons VC and IIIC). Figure 11:5 is a proposed reconstruction of the filling process. Before construction of the raised fields the flats had a brown-black A horizon and a bright black C horizon (here labelled C2) (Figure 11:5a). Ridge construction and agricultural operations destroyed the original A horizon which was mixed with C2 material to produce a slightly lighter black loam on the ridges (here labelled C1) Figure 11:5b). Most of the ditches probably contained standing or flowing water for subirrigation, frost amelioration, and accumulation of silt and muck for fertilization. Organic material enriched the C2 material at the bottom of the ditch, creating a new anthropogenic A horizon (VIA) (5 percent organic matter, see Appendix).

A layer of white volcanic ash (layer V) was deposited on the landscape and was blown and washed from the ridges into the ditches (Figure 11:5c). At this point, the raised fields were abandoned. We know they were not abandoned earlier since there is no fill beneath the volcanic ash layer. The white ash is low in phosphorus and organic matter (see Appendix). When the ditches were without maintenance, material quickly silted in from the ridges (lay IV, Figure 11:5d) and this material has a higher organic matter content than the adjacent ash materials (see Appendix). In addition, a brown-black A horizon may have begun to develop where the black C1 and IV materials were most stable.

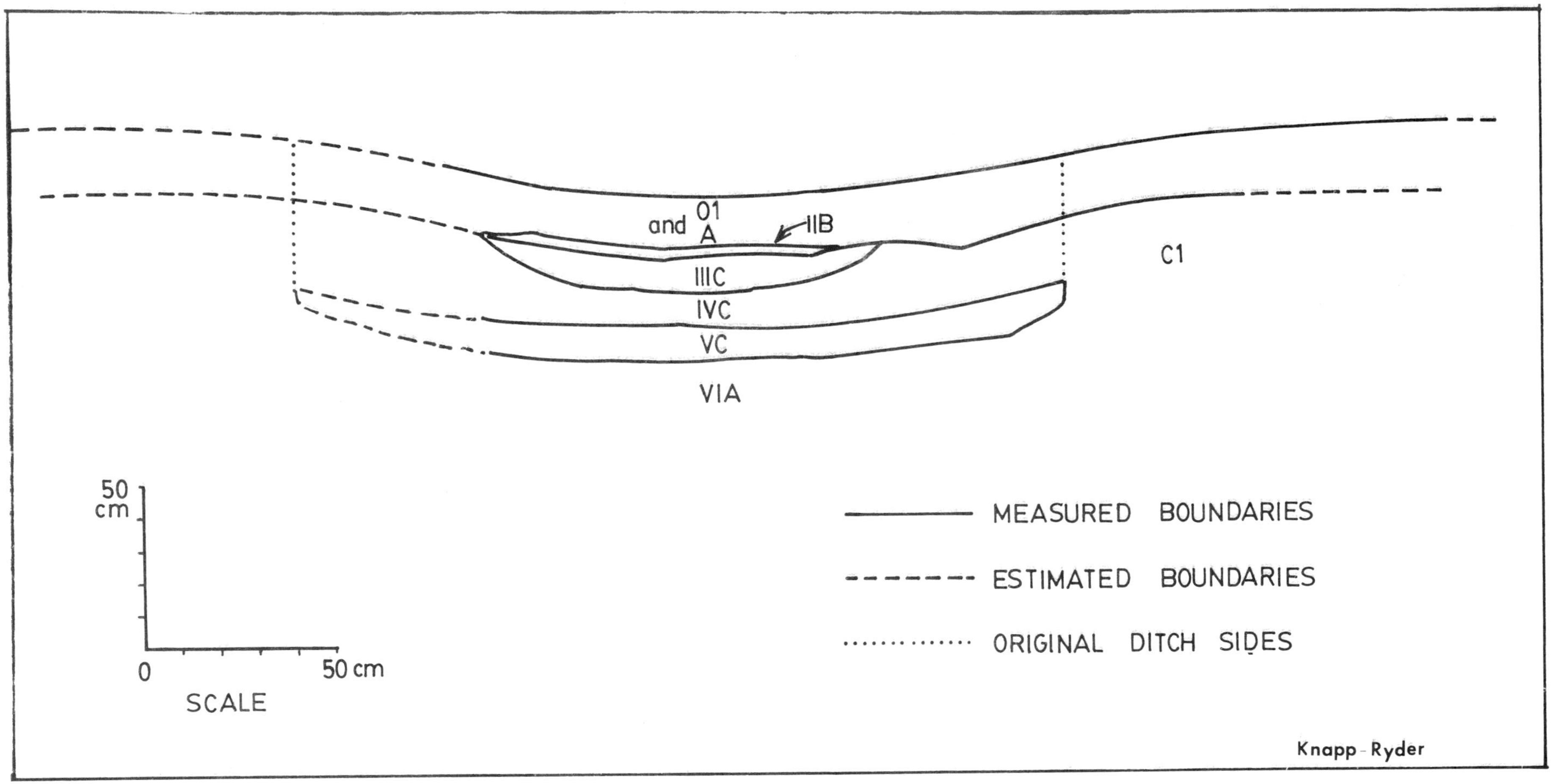


<u>Figure</u> <u>11:4</u> Ditch cross section at Santa Rita ridged field site

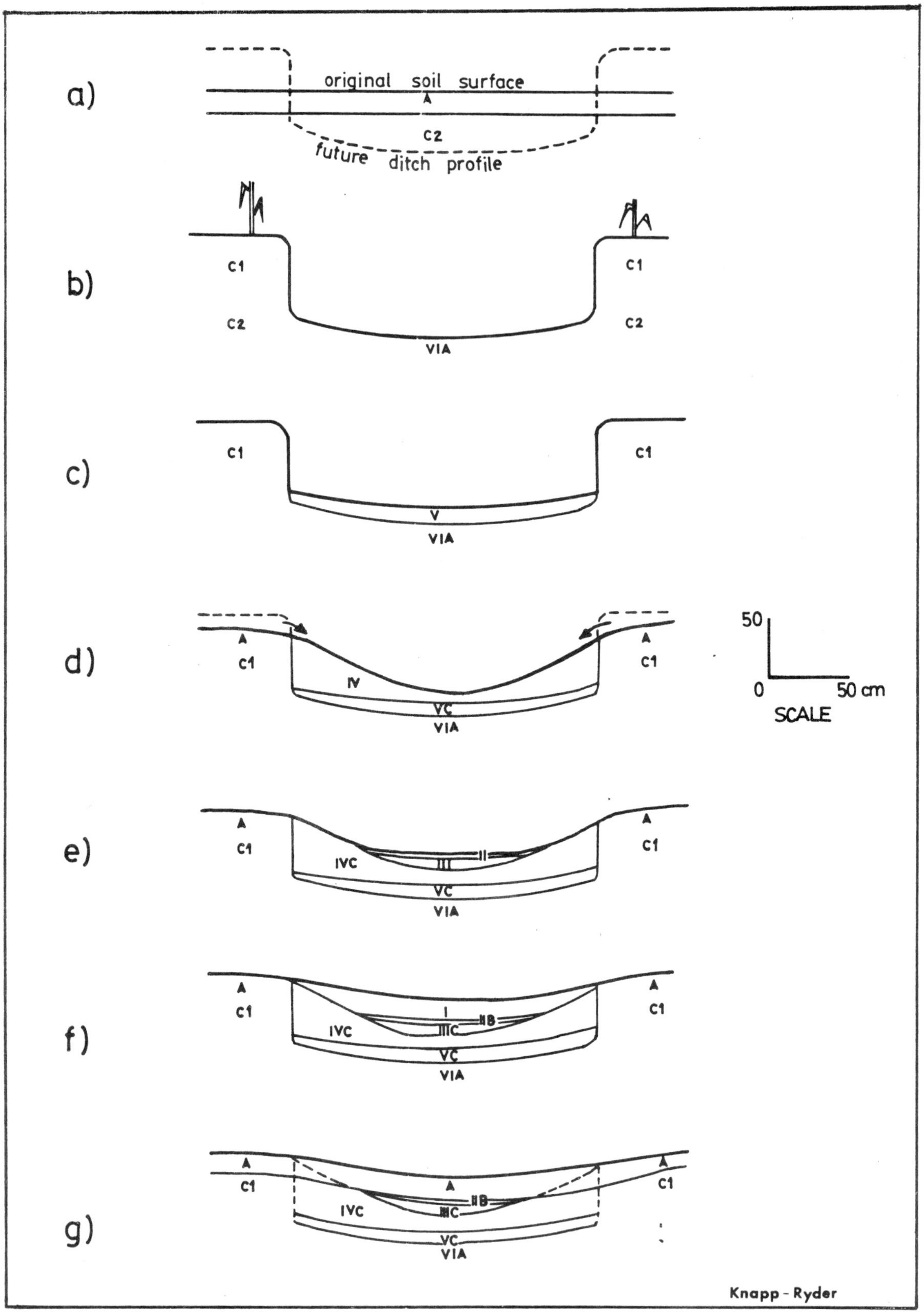


Figure 11:5 The filling process at Santa Rita

Additional layers of volcanic ash (III and II) fell on the plain and were transported into the ditches (Figure 11:5e). With time, these ash layers were capped by material transferred from the ridges to the ditches by erosion and mass movement (I, Figure 11:5f). This layer exhibits the highest phosphorous concentrations of any of the ditch fill horizons; at 34 kilograms/hectare phosphorous, it is also much richer than the adjacent ancient ridge (the latter exhibited 10 kilograms/hectare phosphorous at 50 cm depth). Since phosphorous is the single most important limiting factor for agriculture in these soils, the material accumulating in the ditches between raised fields (always excepting volcanic ash falls) would have been valuable for promoting soil fertility on the planting surfaces.

With more time, the lighter-coloured A horizon which was developing over the entire landscape became more pronounced (Figure 11:5g); dark organic colourants were leached into the subsoil. The fluctuating water table promoted the formation of Fe/Mn concretions in the developing IIB horizon. Iron leached down through root channels. The boundary between horizons I and the adjacent ridges became obscured by the formation of the new A horizon. As the IVC horizon was produced from slumped ridge material, the ditch boundary is also observed with difficulty at the IVC/C1 contact. Only the edges of the first ash deposit (V) indicate clearly the original sides of the ditch. It is impossible at the moment to date either of the two eruptions; successful dating of the first ash fall would provide precise information on the date of field abandonment.

## Discussion: Raised Field Function and Carrying Capacity

Elsewhere Knapp (1983) has provided a discussion of the probable functions, labour imputs and productivity of prehistoric raised fields, based on a variety of evidence from the northern Ecuadorian highlands. Since the evidence from the Santa Rita fields presented in this report is consistent with that from the other sites, the conclusions of Knapp (1983) will be summaried here (see also Knapp and Denevan 1983).

The remarkable depth of the ditches (compared with contemporary potato ridges) and the relatively short wave-length of the raised fields indicate that very high labour inputs were required for ditch cleaning; for example, modern practice for deep ditches involves from one to four cleaning operations per year. Such high labour inputs would only have been justified by a high agricultural productivity. This high productivity would have been faciliated by constant soil renovation on the ridges, sub-irrigation from water in the ditches, and the frost-ameliorating properties of high ridges between fingers of standing water. In view of the preceding discussion of frost risk (Table 11:1), this probably means tht potato double-cropping or potato-maize double-cropping were feasible, as currently practiced in favourable locations. Based on modern ethnographic data, these cropping practices may have enabled over 700 persons to be supported per square kilometre of raised fields. A plausible late prehistoric settlement population of 3000 (Athens 1980: 180) could thus have been supported on 430 hectares of flats, a small portion of the south Quito altiplano.

## Conclusions

The discovery of these raised fields on the Quito altiplano raises to five the number of large poorly drained flats which exhibit such features in highland Ecuador; several small flats have also yielded remains of these features (Ryder 1970; Gondard and Lopez 1981). Coupled with the discovery of similar raised fields in the Bogota altiplano (Broadbent 1968), the highlands of central Peru (Earle *et al*. 1980) and the Titicaca basin (Smith, Denevan and Hamilton, 1968), these findings point increasingly to the importance of poorly drained flats as a key niche in prehistoric Andean economies. Today, used primarily as cattle pasture, such flats in the past were probably the locales of demographic, economic and political cocentration, facilitated by their special appropriateness for agricultural intensification.

The discovery of the Quito raised fields sheds new light on the choice of this site by the Inca for their northern capital, which may have been conditioned by ecological as well as military factors. The presence of volcanic ash in the ditches provides an explanation for the limited preservation of visible ridges in the area as well as a future opportunity for the dating of raised field abandonment.

## Appendix

Soil Profile: Chillogallo Raised Field Furrow

Date: 3 December 1981, 2.00 p.m.

Location: Ecuador, Pichincha Province, south of Quito, Barrio Santa Rita, pasture at the end of Cusubamba Street.

Observers: Gregory Knapp and Roy Ryder.

Site Description:

(a) General: Flat pastures dissected by relic river valley.

(b) Slope: In general, under 3%; at site, 5%, in direction of long axis of ridges.

(c) Topographic position: Flat fluvial terrace above slightly incised relic river valley.

(d) Geology: Fluvio-lacustrine and wind-transported deposits of volcanic origin (mostly volcanic ash).

(e) Water table: now deeper than 1.6m; in the past it was close to the surface and was associated with marsh vegetation (local informant). Artificial drainage is relatively recent (30-40 years).

(f) Erosion: Slight due to gentle slope; the raised field furrows have been filled with material slumped and eroded from the ridges.

(g) Elevation: 2855m.

(h) Climatic characteristics: Corresponds to the nearby Quito Observatory. Mean temperature 13.1°C; Annual precipitation 1222 mm; Frost risk in all months of the year; Lowest recorded temperature at surface -4.9°C, November 1948.

(i) Vegetation and fauna: Most of the flats, including the ridges, furrows and the relic river valley, are covered with a dense mat of _kikuyu_ (_Pennisetum clandestinum_), a grass introduced to Ecuador from Africa by way of Colombia in the 1940's. There are isolated hedgerows of Eucalyptus, _chilca_ (_Baccharis_ sp.), etc. Standing water in ditches associated with such plants as _berro_ (_Nasurtium oficinale_). Pastures now used for cattle grazing. Earthworms and grubs in the soil.

(j) Probable environmental history: Colonial records indicate that the flats in this area were used by the Spanish townsmen living in Quito. The flats were described as brushy and marshy by Rodriguez de Aguayo.

Description of Horizons:

| Horizon | Depth (cm) | Description |
|---|---|---|
| O1 | 5 - 0 | A compact mass of roots andpartially decomposed leaves and roots. |
| A | 0 - 11 | Light brownish grey (10YR 6/2) dry, dark brown (7.5YR 3/2) moist, silt loam; moderate fine to medium subangular blocky structure; friable moist; many fine roots; earthworms and earthworm casts present; dark organs on some peds; pH 6.2; abrupt smooth boundary. Phosphorus test: 34 kilograms/hectare. O.M. 6.9%. |
| IIB | 11 - 14 | Light brownish grey (10YR 6/2) dry, dark brown (7.5YR 3/2) moist, coarse sand with many gravel-sized Fe/Mn particles (probably concretions); |

| | | |
|---|---|---|
| | | single-grained structure; many roots; abrupt wavy boundary. |
| IIIC | 14 - 24 | Light grey (10YR 7/1) dry, very dark grey (10YR 3/1) moist, loam to sandy loam; ferrans on side of root channels; some gravel-sized inclusions of material resembling A, IVC and VC; massive structure; very abrupt, wavy boundary. Phosphorus test: 12 kilograms/hectares. O.M. 2.2%. |
| IVC | 24 - 35 | Grey (10YR 5/1) dry, black (10YR 2/1) moist, loam to silt loam, some gravel-sized inclusions of material resembling IIIC and VC; massive structure; pH 6.4; abrupt wavy boundary. Phosphorus test: 9 kilograms/hectare. O.M. 3.9%. |
| VC | 35 - 45 | White (10YR 8/1) dry, light grey (10YR 7/1) moist, loam; slight iron staining on root channels; massive structure; abrupt smooth boundary. Phosphorus test: 9 kilograms/hectare. O.M. 0.4%. |
| VIA | 45 - 73 | Very dark grey (10YR 2/0) moist, silt loam; moderte medium subangular blocky breaking to moderate miedium granular structure; many fine roots; pH 5.8. phosphorus test: 6 kilograms/hectare. O.M. 5.0%. |


Acknowledgements

This project was funded by Fulbright and National Science Foundation grants (NSF grant number 144R345), and carried out with a permit from the National Institute of Cultural Patrimony ad a contract with the Otavalo Institute of Anthropology. The authors also received valuable assistance from the Museum of the Central Bank of Ecuador, the Panamerican Center for Geographic STudies and Invesitgations (CEPEIGE), the National Program of Agrarian Regionalization (PRONAREG-ORSTOM), Maria del Carmen Molestina, John Isaacson, Minard Hall, Hernan Crespo Toral and William M. Denevan.


Bibliography


ANONYMOUS 1965 (1573), La Cibdad de Sant Francisco del Quito, In, Relaciones Geograficas de Indis, edited by Marcos Jimenez de la Espada, Biblioteca de Autores Espanoles, 184, Madrid, vol.2, 205-232.

ATHENS II, J.S. 1980, El Proceso Evolutivo en las Sociedades Complejas y la Ocupacion del Periodo Tardio-Cara en los Andes Septentrionales del Ecuador, Instituto Otavaleno de Antropologia, Otavalo, Ecuador, 307pp.

BATCHELOR, B. 1980, Los camellones de Cayambe en la Sierra de Ecuador, America Indigena 40, 671-689, Mexico.

BROADBENT, S.M. 1968, A prehistoric field system in Chibcha territory, Colombia, Nawpa Pacha 6, 135-147.

CALDAS, F.J. 1942, Bosquejos Ineditos, Seminario del Nuevo Reino de Granada, Vol.3, Re-edited by Biblioteca Popular de Cultura Colombiana, Historia 9, Editorial Kelly, Bogota.

DENEVAN, W.M. 1970, Aboriginal drained-field cultivation in the Americas, Science 169, 647-654.

DENEVAN, W.M. 1980, Tipologia de configuraciones agricolas prehispanicas, America Indigena 40, 619-652.

EARLE, T.K., T.N. D'ALTROY, C.J. LeBLANC, C.A. HASTORF and T.Y. LeVINE 1980, Changing settlement patterns in the upper Mantaro Valley, Peru, Journal of New World Archaeology 4, 1-49.

ECHEVERRIA J.H. 1976, Contribucion al Conocimiento Arqueologico de la Provincia de Pichincha: Sitios Chilibulu y Chillogallo, Pontifica Universidad Catolica del Ecuador, Facultad de Ciencias de la Educacion, Quito.

FREI, E. 1957, Informe al Gobierno del Ecuador sobre Reconocimientos Edafologicos Exploratorios, Mimeographed report, Quito (FAO office).

FRERE, M., J. REA and J.Q. RIJKS 1975, Estudio Agroclimatologico de la Zona Andina, FAO, Rome.

GONDARD, P. and F LOPEZ 1981, Inventaire Archaeologique Preliminaire des Andes Septentrionales de l'Equateur, PRONAREG-ORSTROM, Quito.

KNAPP, G. 1980, Cambios en el clima de los Andes Ecuatorianos: una tentativa explicacion genetica, Revista del Centro Panamericano de Estudios e Investigaciones Geograficas, Quito, December, 39-48.

KNAPP, G. 1981, El nicho ecologico llaura humeda en la economia prehistorica de los Andes de altura, Evidencia etnohistorica, geografica y arqueologica, Sarance (Ecuador) 9, 83-95.

KNAPP, G. 1983, Soil, Slope and Water in the Equatorial Andes: a Study of Prehistoric Agricultural Adaptation, Unpublished Ph.D. dissertation, University of Wisconsin, Madison, 349pp.

PRONAREG-ORSTOM 1980, Mapas de Suelos por Regionalizacion, Second revision, Quito and Martinique.

RODRIGUEZ DE AGUAYO, P. 1965, Descripcion de la Ciudad de Quito y vecindad de ella por el arcediano de su iglesia, licenciado Petrod Rodriguez de Aguaya, In, Relacioes Geograficas de Indias, edited by Marcos Jimenez de la Espada, Biblioteca de Autores Espanoles, 184, Madrid, vol.2, 201-204.

RYDER, R. 1970, El valor de la fotografia aerea en los estudios historicos y arquelogicos del Ecuador, Revista Geografica del Instituto Geografico Militar 6, 40-42, Quito.

SALOMON, F. 1980, Los Senores Etnicos de Quito en la Epoca de los Incas, Instituto Otavaleno de Antropologia, Otavalo, Ecuador, 370pp.

SMITH, C.T., W.M. DENEVAN and P. HAMILTON 1968, Ancient ridged fields in the region of Lake Titicaca, Geographical Journal 134, 353-367.

UHLE, M. 1954 (1923), The aims and results of archaeology, In, Max Uhle, 1856-1944, edited by John Rowe, University of California Publications in American Archaeology and Ethnography, Berkeley, 54-100.

## Chapter 12

# PREHISTORIC INTENSIVE AGRICULTURE: PRELIMINARY NOTES ON RIVER CANALIZATION IN THE SACRED VALLEY OF THE INCAS

Ian S. Farrington


### Abstract

River canalization was an effective means of land reclamation during the prehistoric period in the Central Andes. In the Sacred Valley of the Incas over 23 kms of the Vilcanota River between Pisaq and Chilca are canalized in at least 7 separate schemes. Each is associated with the straightening of the lower reaches of tributary streams and some tributaries are also canalized at higher altitudes. The reasons for canalization are mainly drainage, flood control and the prevention of meandering in order to effectively reclaim land for maize and potatoes. Some schemes were probably built for aesthetic or religious reasons. Documentary evidence indicates that canalization schemes were built by the Incas to reclaim unused land near Coya and although an Inca date is often given for all schemes evidence is presented which casts doubt on that conclusion.



### Resumen

En tiempos prehistóricos, probablemente durante el período Inca, fueron canalizados más de 23 Kilómetros del río Vilcanota, entre Pisaq y Chilca. Además, en la misma área, también fueron canalizados numerosos tributarios al cruzar conos aluviales y llanuras irrigables. El presente trabajo ofrece la información técnica sobre la construcción de ellos, y sugiere las razones que la indujeron:

1) para utilizar tierras para el cultivo que durante la época de cultivo eran inadecuadas;
2) para enderezar los ríos, reduciendo los meandros, para incorporar campos más grandes;
3) para mejorar y utilizar tierras baldías, con propósitos tanto estatales como privados;
4) por razones estéticas o religiosas.

El trabajo presentará ejemplos de cada caso, aunque la canalización del río se debió probablemente a la combinación de varias de las razones descritas.


The study of prehistoric agricultural systems in the Central Andes of Peru has focused almost entirely on the description of hillside terracing (Cook 1916; Donkin 1979), and on raised fields in highland basins (Smith, Denevan and Hamilton 1968; Matos 1978; Parsons 1981; Lennon 1982). Few studies have been undertaken to view the range of agricultural systems currently practised in a small area and the prehistoric field evidence for these (Farrington 1978). One type of prehistoric agricultural construction remains virtually neglected, namely river canalization, although it played a major and spectacular role in the reclamation for cultivation of tracts of valley floor, alluvial fan and former lake beds.

References to this type of system are scattered and generally form only a small part of an archaeological description of an area (e.g. Gonzalez and Franco Inojosa 1937) or are included in travellers' accounts (Squier 1878, 488, 507). Only Regal (1970) presents any formal data on such schemes, although he fails to isolate canalization as an important component of the agriculture of this part of the Andes. This paper presents some preliminary notes on research currently being undertaken on various canalization schemes in the Sacred Valley of the Incas[1], i.e. the Rio Vilcanota between Pisaq and Chilca (Figure 12:1). It will attempt to demonstrate the effectiveness of canalization in draining lands and suggest other reasons for such costly undertakings.

The Sierra in the Department of Cusco can be divided into a series of agro-ecological regions on the basis of altitude, climate and crop complex (Gade 1975; Farrington 1978; see also Table 12:1). The integration of these into the economy of communities and states, both prehistoric and contemporary, has been demonstrated clearly with the concept of the 'vertical economy' (Murra 1972; Brush 1976; Farrington 1978).

The Sacred Valley lies within one of these regions, namely the Quechua, or principal maize producing region (2400-3300m). It is a relatively gently sloping stretch of the Rio Vilcanota, over 70 kms long and ranging in width from ca. 0.5-2 kms. Only downstream of Pachar does it narrow considerably. The valley floor comprises flood plains, river terraces and the alluvial fans of tributary streams and thus presents a variety of microenvironments. The flood plain itself, where it is not cultivated comprises three distinct zones: marshes, highly salinised areas, and porous river gravels. The valley sides are very steep and geomorphically active, yet support much agriculture. The valley is flanked by the Vilcanota range to the north which provides very steep and generally poorly productive lands, but to the south are the high rolling plains of Chita, Anta and Maras (3300-3800m). It is the latter which demonstrates the traditional pattern of agriculture in the area. It is significant though that this area contains very few remains of prehistoric agricultural activity, such as terracing.

In contrast the Sacred Valley has many extensive areas of terracing and river canalization. These are found mainly on the lower slopes, alluvial fans and flood plains. Therefore each of the valley microenvironments was exploited in prehistory by different types of system. The relatively free-draining slopes and fans have been used for millenia and only partially transformed into terraces. The floodplain was also reclaimed in part by canalization.

Over 23 kms of the Vilcanota, between Pisaq and Chilca, was canalized in prehistory in at least six separate schemes; Pisaq, Coya, Calca, Paucarbamba-Pachar,

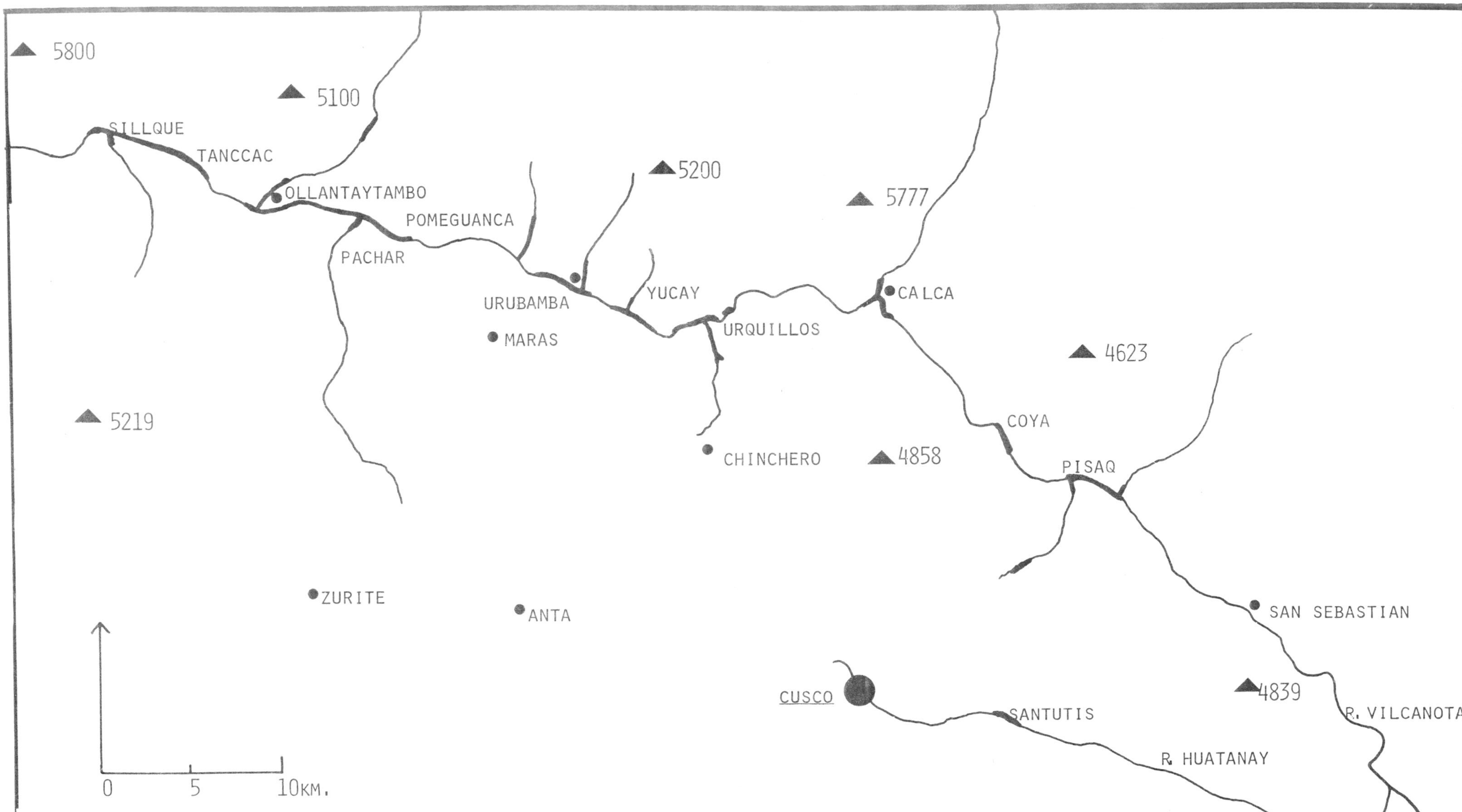


**Figure 12:1** Sacred Valley of the Incas showing the canalization schemes

| INCA | GADE (1975) | CUSICHACA |
|---|---|---|
| PUNA | Native wild grass (pasture) and cultivated potatoes<br>4310-3910m | 1. Punas and pastures<br>4000m +<br><br>2. Main crop potatoes<br>4000-3700m |
| SUNI | Wheat, barley, broad beans, potatoes. Also native tubers, legumes, pseudo-cereals<br>3910-3300m | 3. Maway<br>3700-3200m<br><br>4. Seed/root transition zone<br>ca. 3100m |
| QUECHUA | Maize<br>3300-2400m | 5. Hatun, Tarpuy, maize<br>3000-2800m<br><br>6. Chakimayo<br>2800-2700m<br><br>7. Lower Valley<br>2600-2400m |
| CHAUPIYUNGA | Subtropical starches<br>2400-1500m | |
| YUNGA | Cocoa and coffee<br>1500-1000m | |
| RUPA RUPA | Manioc and cocao<br>1000-750m | |

Table 12:1 Agro-ecological zones of the Central Andes and sub-zones of Cusichaca

Pachar-Ollantaytambo, Tanccac-Sillque. There are traces of a seventh at Yucay, where meandering has been considerably reduced to reclaim the floodplains of Antapacha and Huachac. Also there are several short, 100-200m lengths of straight and walled river bank. Some evidence suggests that the river around Ollantaytambo was intended to be completely canalized but the evidence on the Hacienda Cachicata is equivocal (Table 12:2).

Each of these schemes is associated with the canalization of the lower reaches of tributary streams. For example, at Pisaq the Chongo and Pinchimayoc are both canalized for at least 600m as they cross the flood plain to join the main river (Figure 12:2). At Coya there has been much reorganization of the lower Quebrada Huaynapata in recent times and evidence for its canalization between the town and the river no longer exists. Above the town the stream was straightened for a short distance. Each of the main quebradas in the Yucay area was also canalized. This strengthens the case for the control of the main river here too. In contrast the Rio Phiri which enters the Tanccac-Sillque scheme on the northern bank was not canalized, nor was the Qba. Cachicata.

In addition several tributary streams have also been canalized at higher altitudes. For example, the Patacancha above Ollantaytambo was straightened for a distance of 2.325 kms at ca. 3300m and is associated with a small group of flood plain terraces. The upper reaches of the two main tributaries of the Cusichaca river, the Pampacahuana and the Quesca, are also canalized at Hatunpampa and Chchurupampa respectively at altitudes over 3600m. Each of these systems have small, ca. 0.8m wide, tributary drains flowing into them. Thus it can be seen that canalization occurs at various altitudes within both the main maize and potato growing regions. The distribution of canalized rivers is not restricted to the Sacred Valley and its tributaries. Other schemes have been recorded at Santutis on the Huatanay to the east of Cusco, and in the Limatambo and Paucartambo valleys.

There are two basic types of construction used in the retaining walls of a canalized river. The first involves the building of a 2-8m wide embankment on both banks of the proposed river course. This is usually made of large boulders, river cobbles and gravel with a steeply battered or vertical fitted stone facing wall on both sides. The facing stones range from as little as 0.5m to over 10m in diameter. The embankment may be terraced to give greater strength to a high structure, such as at Pisaq, or it may be a simple double wall about 2m in width, such as in the tributaries flowing into the Vilcanota at Pisaq and in the Cusichaca examples (Figure 12:2). The height of the wall appears to have been sufficient to prevent flooding during an average summer. However the cross sections through the northern bank of the Vilcanota at Pisaq indicate that occasional overbank flooding did occur. Alongside each of the tributary schemes a paved path was also built as part

| Scheme | River | No. of tributaries | Altitude | Length | Width | Slope | Function | Area drained |
|---|---|---|---|---|---|---|---|---|
| Pisaq | Vilcanota | 2 | c.2960 | 2.9 kms | c.70m | 0.002 | drainage | 183 ha |
| | Chongo | | | 0.7 kms | 7m | 0.02 | | |
| | Pinchinmayoc | | | 0.6 kms | 5m | 0.013 | | |
| Coya | Vilcanota | 1 | c.2935 | 0.65 kms | 50m | 0.002 | drainage | 23 ha |
| Calca | Vilcanota | 1 | c.2918 | 2.25 kms | 50m | 0.003 | drainage | 169 ha |
| | Ccochocc | | | 0.75 kms | 5m | 0.022 | | |
| Yucay | Vilcanota | 4 | c.2840 | probably 10 kms | 50m | | | |
| | Urquillos | 1 | c.2900 | 3.4 kms | 3m | 0.015 | drainage | 19 ha |
| | Urpihuata | | | 0.5 kms | 2m | | | |
| | San Juan | 2 canals | c.2900 | 1.0 kms | 1.5m | 0.09 | flood control | |
| | Chicon | | c.2950 | 2.4 kms | 2m | 0.062 | flood control | |
| | Pomehuanca | 2 canals | c.2940 | 1.4 kms | 1m | 0.064 | flood control | |
| Ollantaytambo | | | | | | | | |
| a. Paucarbamba - Pachar | Vilcanota | 1 | c.2820 | 3.25 kms | 40m | 0.002 | drainage | 97 ha |
| | Pachar | | c.2830 | 0.9 kms | 8m | 0.023 | drainage | 16.5 ha |
| b. Pachar - Ollantan-taytambo | Vilcanota | 1 | c.2800 | 8.4 kms | 50m | 0.031 | meander control drainage | 40 ha |
| | Patacancha | | c.2840 | 1.5 kms | 8m | | | |
| c. Cachicata | Vilcanota | 0 | c.2780 | possible | | | | |
| d. Tanccac Sillque | Vilcanota | 1 | c.2770 | 5.45 kms | 60m | 0.018 | drainage | 228 ha |
| | Sillque | | | 0.55 kms | 6m | 0.04 | drainage | |

Table 12:2 River Canalisation Schemes of the Sacred Valley

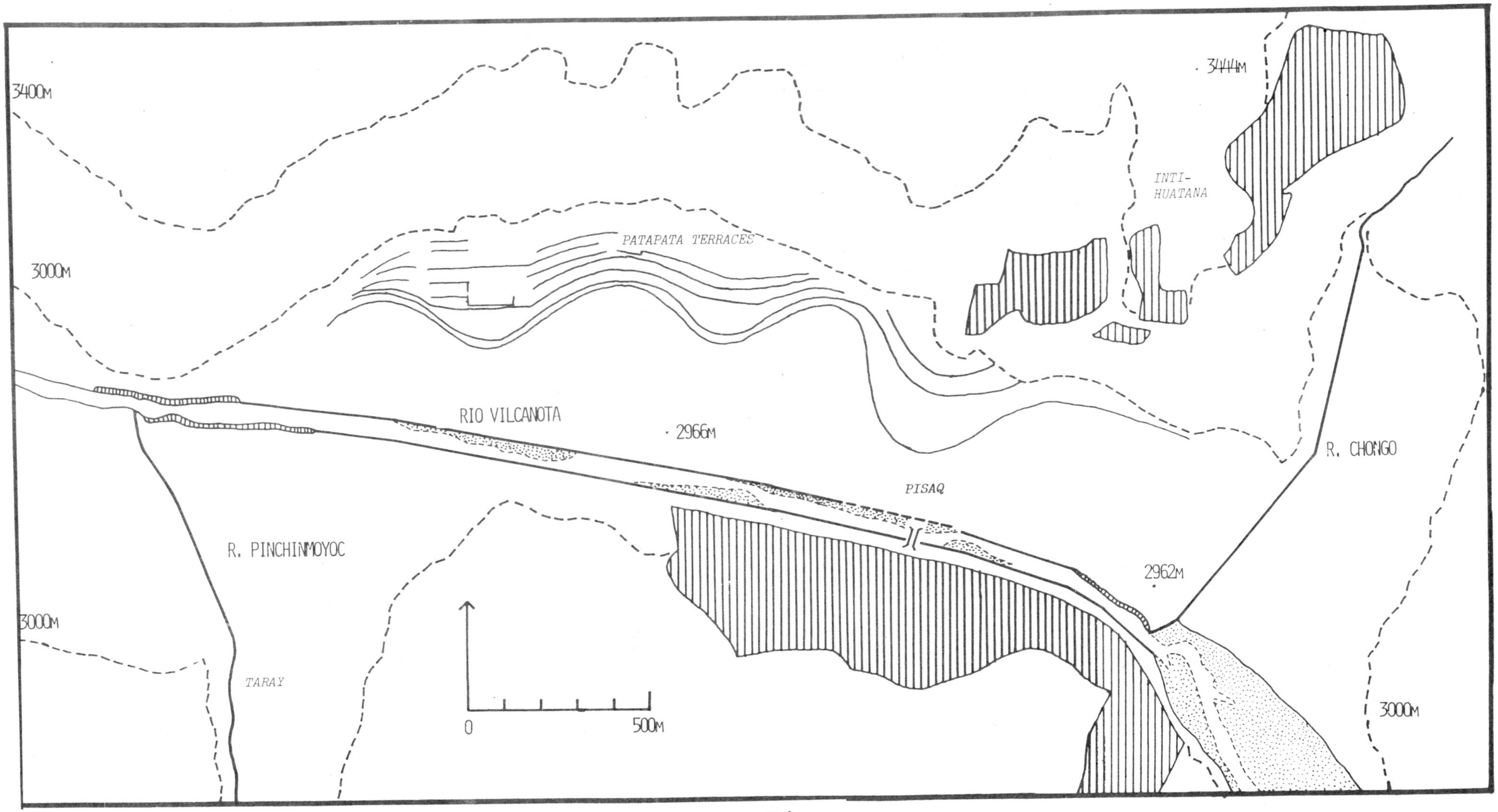


Figure 12:2 Canalization of the Vilcanota River at Pisaq

of the overall plan.

The second type is found where the river is incised. Its banks are faced with a well-fitted stone wall to a height of 3-4m and on top of this there may be a further 2-4m wide retaining wall, such as near Pachar. A variaton of this type occurs when the retaining wall is merely the lowest of a flight of narrow terraces, such as near Pisonayniyoc in the Pachar-Ollantaytambo scheme. In some places one bank may be of the first type with a high embankment whilst the other may be of the second.

Major construction work must have taken place during the winter months when the river was at its lowest and in places was even fordable. At this time it would have been easy to redirect flow or channel it into a narrower course in the middle of the proposed man-made canal. It is not known which of these two possibilities was practised.

An obvious feature of canalization schemes is their straightness, although in some schemes curves were deliberately constructed. The location of curves was either dictated by the geography of the valley, such as in the Patacancha, or where the maximum discharge was so large that a very long straight would have been almost impossible to maintain. Between Pachar and Ollantaytambo curves were deliberately built with very sharp angles. On the outside of these partial weirs were built to disrupt the normal flow pattern and thus reduce scouring of the outer bank which in this length were the lower walls of a flight of terraces. Gentle curves were built into many systems, such as the long Tanccac-Sillque canalization, but these must be regarded as hydraulically straight. It is interesting to note that this scheme which carries the largest discharge does not have any evidence of bank erosion.

The confinement of a river with a discharge as large as the Vilcanota (average summer discharge at Pisaq is about 100 $m^3$ $sec^{-1}$) into a straight or hydraulically straight channel is a major engineering achievement, but it is not without its problems. Meandering is the normal state for any water channel and a river will always adopt a corkscrew-like flow even in a straight channel. This places great strain on the banks and calls for very strong walls or embankments to avoid erosion and bank collapse. The constructors of these schemes in the Sacred Valley appear to have been well aware of this problem because most of their schemes have survived intact. However breakdowns can be noted. For example, near Ollantaytambo the canalization clearly represents two phases of construction and meandering was responsible for this disruption of the original channel. The reconstruction of the scheme also occurred during prehistory and the scheme now has opposing banks on different alignments with a sharp angle in the northern one to return it to the original course. On the southern bank the original wall survives but a large sand bar has been created as the river pushed northwards in its meandering (Figure 12:3). Most meanders appear to be of more recent date and are probably the result of a lack of maintenance. The Yucay scheme has been almost completely destroyed, and at Hatunpampa marshy ground has been recreated downstream of a prominent meander.

Erosion has also destroyed the original funnel-shaped entrances to most of the schemes. At Pisaq and at Santutis one of the entrance walls survive. Landslides are a perennial hazard in the Andes and one scheme at Chchurupampa in the upper reaches of the Cusichaca has been pushed out of its channel by an enormous landslide.

Sedimentation has occurred in schemes, such as at Pisaq, where the river gradient is gentle and the velocity not great enough to prevent or to cause scour in the channel in the winter months. There the river flows in a slightly narrower channel than that planned. This may have been deliberate policy to allow the river to establish its own stable channel within the confines of a man-made course. This practice is well known in other parts of the world and was used by British engineers in the nineteenth century to canalize the distributaries of the Indus (Blench 1957). In this way the river could be allowed to create its own channel and yet still perform its planned functions very efficiently.

There are a number of physical and cultural reasons for the canalization of the Vilcanota and its tributaries. The main physical one is for drainage, although flood control is also important if fields are to be protected from crop loss by inundation and from salinization. Flood control was effected by embankments. Canalization also prevented the spread of water disgorging from a mountain stream and flooding and scouring its alluvial fan. Meander control is a corollary of drainage. It was particularly important in narrow reaches of the valley where there was a lack of reasonably sized areas of bottom land. The straightening of the river had the effect of reclaiming relatively large fields and protecting them from further erosion. The best example of this can be found in the lower Cusichaca at Tunasmocco where canalization has enabled a very regular terrace system to be laid out. Part of this has now been destroyed as the river has begun to meander again.

The principle of constructing a straight or relatively straight channel to replace one which was meandering across a fairly gentle slope is readily understood. It will move water quicker through the system and drain off excess moisture from surrounding fields. The canalization of tributaries and the construction of drains into the main channel facilitates this process and serves to reduce soil moisture to tolerable levels for plant growth. It is interesting to note that in the Sacred Valley many of the canalization schemes are given over to the production of _miska_ or early maize. This crop is planted at the height

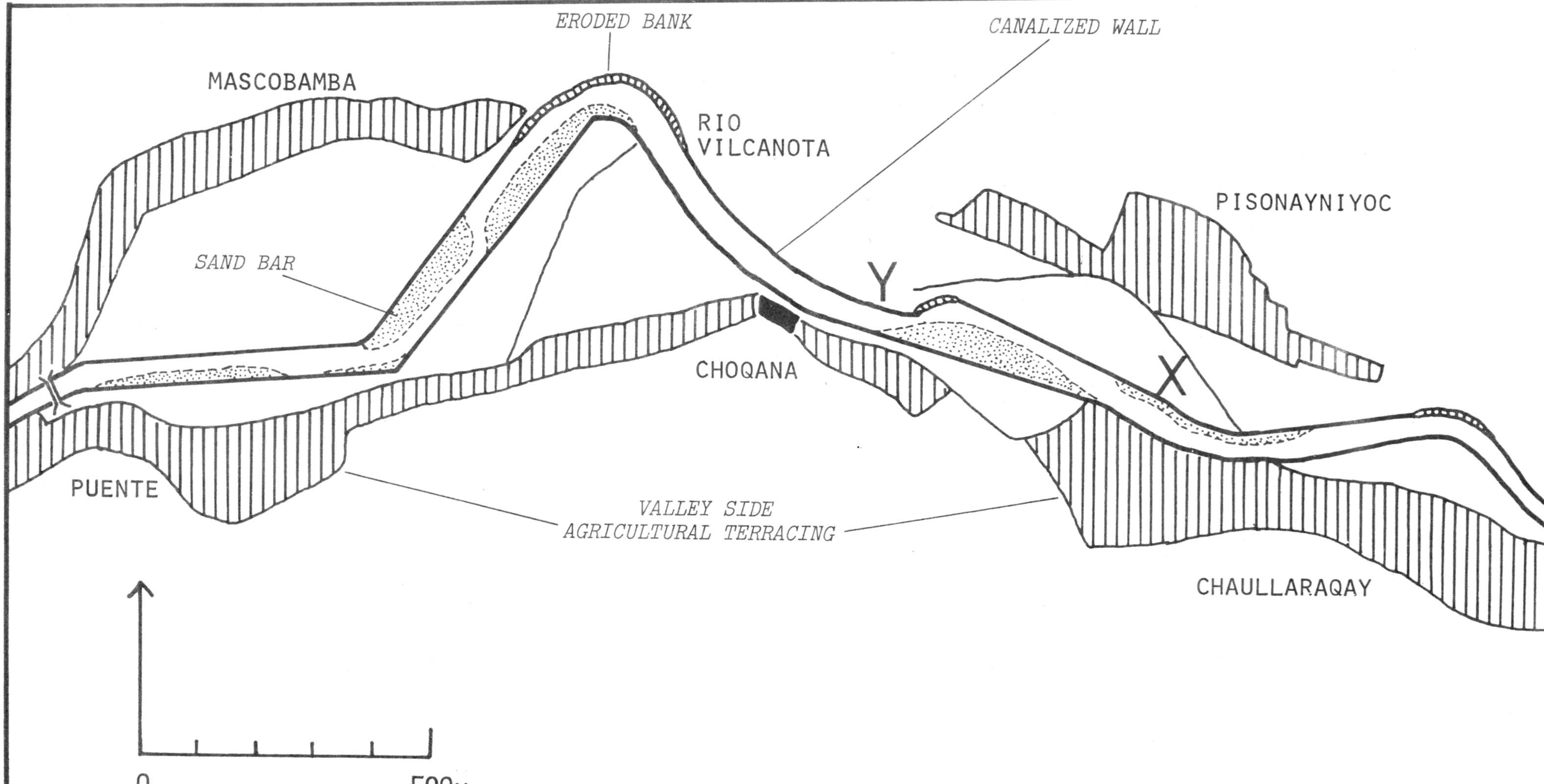


Figure 12:3 Canalization of the Vilcanota River between Pachar and Ollontaytambo. Note the non-parallel walls of the channel between X and Y

of the dry season and in most locations usually requires much irrigation. In the reclaimed flood plain soils the water table still remains relatively high at this time of the year and thus *miska* planted in these areas requires less irrigation. This crop is harvested before the maximum of the summer floods which in some places still renders these soils unproductive. In the puna the canalization of the upper arms of the Cusichaca has enabled two former lake beds to be drained and thus provide rich soils for potato production, although for other reasons one of these schemes no longer functions.

Physical reasons cannot be considered for all the canalization schemes. For example at Ollantaytambo between the bridge and the railway station the river passes between a steep mountainside on its southern bank and the terraced alluvial fan of the Qba. Patacancha to the north and yet its course is walled with good quality stonework. Further downstream in the Cusichaca and Machu Picchu areas the river is also canalized whilst flowing steeply through gorges. The only suggestion that can be offered is that these were built for aesthetic or religious reasons. It is of interest to note that there is a carved rock or *huaca* within 50m of the start of the Tanccac-Sillque scheme. This argument is enhanced when the importance of water in Inca religion and ritual is taken into account (Carrion Cachot 1955; Zuidema 1978).

Drainage and its associated consequences were therefore important considerations to the Prehispanic peoples who built these schemes, but the question to be asked is why build them in the first place. Traditional agriculture in the Sierra takes place in sloping fields on both mountain sides and alluvial fans where available soil moisture is satisfactory for crop growth throughout the season and where the soils are relatively easily worked and yield reasonable harvests. Wet areas are generally avoided unless there is the possibility of producing an early crop, such as *maway* or *miska*, or there is a restriction of the amount of land available within a community.

The Vilcanota flood plain between canalization schemes is characterised by poorly drained and saline soils as well as extensive areas of gravel. It supports relatively little agriculture today and probably even less in Prehispanic times. Therefore there must have been a land shortage in order to stimulate the development of such land reclamation schemes. The nature of this land shortage has to be viewed not only in demographic terms but also in political and economic ones. I would suggest that the construction of such projects was not beyond the capabilities of a community or group of communities. However, the small area reclaimed and as a consequence the small harvest yield probably would not have warranted such an investment from a community. Therefore the schemes must have been constructed to satisfy requirements beyond the needs of such a group and these must be viewed in terms of a socio-political reconstruction of certain periods in prehistory.

In the literature, river canalization is always referred to as a product of the Inca state (Regal 1970; Gasparini and Margolies 1980). Indeed the Spanish chronicler Betanzos, writing in 1551 (1980), attributed the canalization of the rivers through the Inca capital of Cusco to Pachacuti, the ninth Inca king. The major schemes at Pisaq and Ollantaytambo, as well as Yucay, have stonework which is generally regarded as Inca in construction. Furthermore each of these areas has many Inca sites. There is also further chronicle evidence from Sarmiento (1943) which relates the activities of certain Incas to the development of these areas.

Rostworowski (1962) has published one specific document which refers to river canalization by the Incas. This is a land claim by Martin de Meneses which was lodged before the Cabildo of Cusco in 1555. It was for a piece of land called Pomebamba, over 50 *fanegadas* (ca. 120 ha) in extent, which lay vacant alongside the Vilcanota river. The document includes the testimony of the four *caciques* of the nearby communities of Quilloay, Macay, Oma and Paulu[2]. Their statements are basically the same. Don Diego, the *cacique* of Quilloay stated that Pomebamba was previously a gravelly area which no one wanted to farm, but under orders from the Inca, Topa Cusi Guallpa[3], the river was diverted and land reclaimed. This was planted with maize by *mitimaes* groups brought from the four provinces of the Empire and the harvest was shipped directly to Cusco. Furthermore it was stated that after the Spanish Conquest this land was abandoned and the *mitimaes* workers returned to their homelands.

The significance of this document is manifold. Firstly, this scheme can be identified as that of Coya for three of the four communities: Quillhuay, Macay and Paulu, are found today on either side of it. The Coya scheme must have been more extensive than that which can be recognised in the field today. It probably included one curve and extended downstream of Hacienda Macay. Secondly, the document indicates that the local communities played no part in the construction or operation of the canalization. After the Conquest they displayed no interest in maintaining those lands and were probably only consulted in the land claim because they were adjacent. Thirdly, the scheme appears to have been farmed by *mitimaes* labour brought from elsewhere in the empire. Generally these groups were given land for themselves and obliged to farm state lands. The only way the state could obtain land was either by expropriation or by reclamation. It seems therefore that the latter option was adopted in this instance and thus prevented much hostility from the surrounding communities. It is interesting to note that elsewhere in the Sacred Valley agricultural terracing is merely an addition on the flanks of already farmed alluvial terraces. It is only in areas such as Yucay and Ollantaytambo, that extensive lands were expropriated and presumably part of their population removed to make way for state fields, private estates and _mitimaes_ and _yana_ groups to work them. It is suggested therefore that it was the state's need for new lands to support itself beyond that gained from tribute which was the stimulus for land reclamation and hence canalization. This would give the Inca access to productive valley bottom lands in which _miska_ maize, a crop important to Inca ritual, could be grown easily.

It would be easy to suggest on the strength of this evidence that all schemes were built during the Inca period. However near Ollantaytambo it is clear that the original canalization was destroyed and rebuilt on a slightly different alignment during prehistory (Figure 12:3). This could have been during the Inca period but MacDonald has indicated that there are at least two clear phases both in the construction and alignment of agricultural terracing and in the architecture in the vicinity of this reconstruction. What this means is not fully understood but it is certain that we are dealing with a period much longer than the generally accepted 90 years of Imperial Inca domination of the valley. There are two possible answers: that the Inca period must be lengthened or that an earlier state built the first canalized rivers in this part of the valley.

The river canalizations of the Sacred Valley of the Incas were certainly magnificent engineering achievements. That they survive is testimony to the skills of the prehistoric engineers. Their principle purpose appears to have been to reclaim by drainage new lands for the state, mainly for the production of maize. Each major scheme is associated with canalized tributaries to further protect the newly gained lands. There are two relatively standard types of bank construction, which also assist in meander control and flood control. It is proposed that for certain schemes aesthetics and ritual must have been important reasons for their construction. They are generally believed to have been built during the Inca period and there is much evidence to support this, although there is the possibility of earlier construction. This paper has presented some data and interpretations of the river canalization schemes in the Vilcanota valley. The distribution of such throughout the Central Andes is unknown and invites further study.

_Footnotes_

1. The 1982 Mision Arqueologica Australiana to Peru is investigating agricultural terracing and canalization in the Sacred Valley.

2. Extract from the testimony of Don Diego, Cacique of Quillhuay, Archivo Historico Nacional de Lima, Judicial Civil - Legaje Num.2 1955 (Rostworowski 1962:48).

'las cuales dichas tierras ahora se llaman pomebamba y que en tiempos pasados no sabe como se solia llamar y sabe que por dicho pedazo de tierra erial solia ir el rio que agora va por mas abajo y que tode ello era un cascajal y cosa perdida que nadie se aprovechaba de ello y que por mandate del Ynga senor de esta tierra que el dicho Ynga se llamaba Topa Cusi Guallpa se quito el rio de elli y hecho por donde ahora va y hicieron muchas yndios de las cuatro provincias los cuales dichos yndios de las dichas provincias por mandato del dicho ynga despues de haber hechado el rio de alli hicieron chacra y de alli adelante se sembraba maiz para el Ynga y lo que se cogia se llevaba al Cuzco de manera que entonces fue la primera vez que se sembro y rrompio el dicho pedazo de tierra no ha visto que los yndios que tiene dicho de las dichas provincias que lo sembraron para el dicho Ynga en esta tierra/venido los espanoles a ella se fueron los dichos yndios de las dichas provincias a sus tierras y quedaron vacas las dichas tierras ...'.

3. Rostorowski (1962:134) notes that this was either Huayna Capac or Huascar.

Acknowledgements

This research was supported by an Australian Research Grant and was sanctioned by the Nacional de Cultiva, Ministerio de Educacion, Lima.

Bibliography

BETANZOS, J. 1880, Suma y Narracíon de los Incas (1551), Biblioteca Hispano-Ultramarina, vol.5, Madrid.

BLENCH, T. 1957, Regime Behaviour of Canals and Rivers, Butterworth, London.

BRUSH, S.B. 1976, Man's use of an Andean ecosystem, Human Ecology 4, 147-166.

CARRION CACHOT, R. 1955, El culto al agua en el antiguo Peru, Revista del Museo Nacional (Lima), 11, 50-140.

COOK, O.F. 1916, Staircase farms of the Ancients, National Geographic Magazine 29, 474-534.

DONKIN, R.A. 1979, Agricultural Terracing in the Aboriginal New World, Viking Fund Publications in Anthropology Number 56, University of Arizona Press, Tucson.

FARRINGTON, I.S. 1978, Contemporary Agriculture and the Vertical Economy, Land Use and Irrigation Research Report: 2, Cusichaca Project, privately published, Canberra, 65pp.

GADE, D.W. 1975, Plants, Man and the Land in the Vilcanota Valley of Peru, Dr. W. Junk, The Hague, 240pp.

GASPARINI, G. and L. MARGOLIES 1980, Inca Architecture, Indiana University Press, Bloomington and London, 350pp.

GONZALEZ, T.A. and J.M. FRANCO INOJOSA 1937, Los trabajos arqueológicos en el Departamento de Cuzco, Revista del Museo Nacional (Lima), 6, 67-80.

LENNON, T. 1982, Raised fields of Lake Titicaca, Peru: A Pre-hispanic water management system, Ph.D. dissertation, University of Colorado, 340pp.

MATOS, M.R. 1978, The cultural and ecological context of the Mantaro Valley during the Formative Period, In, Advances in Andean Archaeology, edited by D. Browman, Mouton Publishers, The Hague, 307-325.

MURRA, J.V. 1972, El control vertical de un maximo de pisos ecologicos en la economia de las sociedades andenas, In, Vista de la Provincia de Leon de Huanuco en 1562, Inigo Ortiz de Zuniga, Visitador 2, 429-476, Huanuco, Universidad Hermillio Valdizan.

PARSONS, J.R. 1981, Late Prehispanic agricultural intensification in the Unin Puna, Peru, Abstract presented at the conference on Prehistoric Intensive Agriculture in the Tropics, Canberra.

REGAL, A. 1970, Los trabajos hidráulicos del Inca en el antigue Peru, Graf Industrial, Lima, 143pp.

ROSTWOROWSKI DE DIEZ CANSECO, M. 1962, Nuevos datos sobre tenencia de tierras reales en el Incario, Revista del Museo Nacional (Lima), 31, 130-159.

SARMIENTO DE GAMBOA, P. 1943, Historia General Llamada Indica (1572), Biblioteca Emece, Buenos Aires, 302pp.

SQUIER, E.G. 1878, Peru. Incidents of Travel and Exploration in the Land of the Incas, New York.

SMITH, C.T., W.M. DENEVAN and P. HAMILTON 1968, Ancient ridged fields in the region of Lake Titicaca, Geographical Journal 134, 353-367.

ZUIDEMA, R.T. 1978. Lieux sacrés et irrigation: tradition historique, mythes et rituels au Cuzco, Annales ESC 1037-1056.

Chapter 13

# RAISED FIELD COMPLEXES AND ASSOCIATED SETTLEMENTS IN THE COASTAL PLAIN OF WESTERN SURINAME

A.H. Versteeg


## Abstract

Two types of settlements are associated with raised fields in the coastal plain of western Suriname: settlements on the sandy chenier ridges and, in the western part of the region, where ridges are scarce, on artificial clay mounds.

Recent investigations of the mounds have proved that two ceramic styles are represented: one mound yielded pottery of the Mabaruma style and $^{14}C$ datings from c.1850 B.P. onwards, while two westerly-situated mounds yielded pottery of the Early/Late Hertenrits style and datings from c.1200 B.P. onwards. Size and form of the associated raised fields also show differences.

## Resumen

Dos tipos de colonización se encuentran asociados con campos elevados en la llanura costera del oeste de Surinam: colonización sobre las lomeras chenier arenosas y, en la parte occidental de la región donde escasean las lomeras, sobre terraplenes arcillosos artificiales. Investigaciones recientes sobre los terraplenes han revelado la existencia de dos estilos de cerámica: un terraplén contenía cerámica del estilo Mabaruma y fechas de $C^{14}$ a partir de alrededor de 1.850 Antes del Presente, mientras que dos terraplenes ubicados al oeste contenían cerámica del estilo de Comienzos/Fines del Hertenrits y fechas de $C^{14}$ a partir de 1.300 Antes del Presente. El tamaño y la forma de los campos elevados asociados también indican diferencias.


## Introduction

In Suriname - the middle part of the Guianas - archaeological investigations have been scarce and the results have not always drawn the attention they deserve. For instance, recent investigations of the archives of the National Museum of Antiquities at Leiden, The Netherlands, revealed that as early as 1887 a prehistoric habitation mound was discovered in the coastal wetlands of Suriname, in the Coronie District near Burnside. Although the find was reported at the time the existence of habitation mounds in Suriname remained unresearched until the 1950's (Versteeg 1983).

In the 1950's the discovery of several mounds was made in conjunction with the expansion of rice cultivation in the west of the country. Before this agricultural expansion cropping was restricted to the land associated with rivers and creeks. The recent sediments of the Young Coastal Plain of western Suriname offer a good soil for agriculture and research revealed that these wetlands had been used extensively for agricultural purposes in pre-

Columbian times.

Several habitation mounds close to complexes of raised fields have been discovered since 1955. These studies relied heavily on air photographs as a data source. The chenier-associated raised fields have been found predominantly in the eastern part of the chenier plain, and, additionally, in eastern Suriname but also near Peruvia in western Suriname, c.f. Figure 13:1 (Boomert 1976). Remarkably, raised fields have not been found in the central part of the coastal plain of Suriname. It has been argued that patterns of agriculture in colonial times have probably destroyed them (Boomert 1976) but it may be that none ever existed. Raised field patterns are still revealed by the differential growth of the rice crops in the empoldered areas of the rice fields of western Suriname, despite levelling of the land with modern heavy equipment - this indicates that these patterns are not readily destroyed.

## Archaeological Investigations

Archaeological fieldwork in 1957, 1962 and 1970 was concentrated on the largest and highest mound, the Hertenrits, and associated raised fields; additional excavations were undertaken at two other habitation mounds situated near the Hertenrits and at Peruvia and Prins Bernhard Polder (Figure 13:1).

While much data have been collected from the mounds and chenier sites near raised fields, investigations of the raised fields themselves have not been very rewarding up until now, partly because of the scarcity of artefacts and datable material in the fields and partly because of the difficulties of investigating the fields under the prevailing field conditions. The raised fields in western Suriname are covered by recent sediments and they are partly submerged, even during the dry season. Borings indicate that the soil used for constructing the fields is very similar to the subsoil, which makes it difficult to distinguish between the natural and the artificial units. In some fields however, a thin layer of darker coloured humic or peaty material marks the contact between the units.

As complexes of raised fields have been found close to all the mounds, the raised fields are ascribed to the inhabitants of the mounds. The chenier site, Peruvia, situated in the east of western Suriname, yield pottery with several characteristics of the Hertenrits style, like that found in the Hertenrits and Wageningen-1 mounds. A related style, called Barbakoeba, has been found at similar chenier sites in eastern Suriname, again in association with raised fields. However, besides some similarity between the Hertenrits and the Barbakoeba style, many differences are also apparent but these have not been investigated yet.

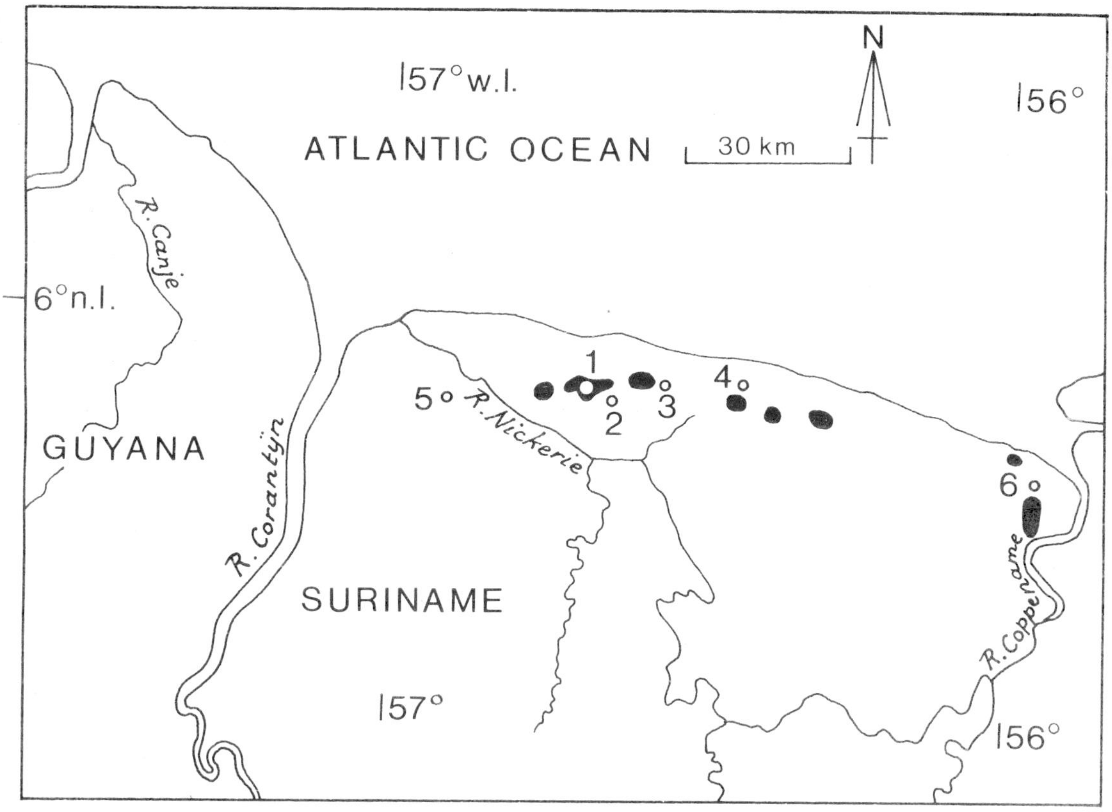


Figure 13:1 Partly after Boomert, 1976, complex of raised fields. 1 Hertenrits mound; 2 Wageningen-1 mound; 3 Buckleburg-1 mound; 4 Burnside mound; 5 Prins Bernhard Polder site; 6 Peruvia site

The Wageningen-1 and Hertenrits Mounds

The Wageningen-1 and Hertenrits mounds are situated at a distance of ca. 4 km from each other on the borders of the same creek system. The 1977/78 investigations at Wageningen-1 and those at Hertenrits, carried out in 1957, 1962 and 1970, indicate that habitation and build-up at both sites started around 1300 B.P. and that both sites were inhabited by the makers of Hertenrits style pottery (Geijskes n.d.; Laeyendecker-Roosenburg 1966; Van Der Heide 1973; Boomert 1978; Versteeg 1980). Both mounds yield pottery of the Early Hertenrits style, characterised by a limited scale of vessel forms and decoration patterns. The upper levels of the Hertenrits mound however, yielded pottery of the Late Hertenrits style characterised by more vessel forms and more elaborate and more varied decoration patterns than the Wageningen-1 mound ceramics. The pottery showed an important Arauguinoid component as recognised by Boomert (1977).

The stratigraphy of the Wageningen-1 and Hertenrits mounds consists of the same three units as the Buckleburg-1 mound to be discussed later (Figure 13:2). It consisted of a marine unit representing the Moleson member, a fresh water unit and the artificial unit. These observations from the sections are confirmed by pollen diagram $H^3$ from the Hertenrits mound (Laeyendecker-Roosenburg 1966). In this diagram (Figure 13:5) the top of the marine blue clay is recorded at a depth of 310 cm. The freshwater unit is seen between a depth of ca. 310 and 220 cm. From ca 220/210 cm and higher; the artificial unit of the mound is indicated by the increase of pollen of Avicennia spp. The results of $C^{14}$ dating show a striking conformity between the Wageningen-1 and Hertenrits mounds; the lowest levels, or unit 2, of both mounds are dated ca. 1300-1250 B.P. at depths of 225-300 cm below the surface. Levels at 60-75 cm yield dates ca. 1150 B.P. Regrettably no dates have been obtained for the highest levels of the Hertenrits mound, from which the pottery of the Late Hertenrits style was recovered (Table 13:1).

A non-mound site situated to the west of the Hertenrits, at Prins Berhnard Polder, was excavated too, and this only contained ceramics of the Late Hertenrits sytle. $C^{14}$ datings of a archaeologically sterile peat layer under the site gave a *terminus post quem* for the site of ca. 950 B.P. In addition, a wooden artefact dated at ca. 800 B.P. came from this site. These datings are considered to be indicative of the last phase of habitation at the Hertenrits mounds.

Wageningen-1 did not yield any pottery of the Late Hertenrits style. This site was probably abandoned before the commencement of manufacture of Late Hertenrits style ceramics.

Only a little evidence indicates the end of habitation at the Hertenrits mound. No artefacts

Table 13:1 List of $^{14}C$ datings

| Site | Pit, Depth Below Surface | Material | Age (conv. y. B.P.) | Lab. No. | References |
|---|---|---|---|---|---|
| Buckleburg-1 | pit 1 190 cm | wood | 1835 ± 45 | GrN-9804 | Versteeg, 1980a |
| | pit 1 175-178 cm | charcoal | 1735 ± 35 | GrN-10304 | Fig. 13:2 |
| | pit 1 55 cm | charcoal | 1315 ± 55 | GrN-9803 | Versteg, 1980a |
| Hertenrits | pit 3 250 cm | wood | 1265 ± 60 | GrN-1897 | Vogel & Waterbolk, 1964 |
| | pit 1 | peaty material with charcoal or charred plant remains | 1130 ± 45 | GrN-1898 | Vogel & Waterbolk, 1964 |
| Peruvia | pit 1 125 cm | charcoal | 1210 ± 30 | GrN-9361 | Versteeg, 1980a |
| Prins Bernhard Polder | 60-65 cm | peat | 935 ± 50 | GrN-8954 | Unpublished |
| | stray find | wood | 790 ± 25 | GrN-9801 | Versteeg, 1980a |
| Wageningen-1 | pit 1 225 cm | wood | 1245 ± 25 | GrN-9061 | Versteeg, 1980a |
| | pit 1 200 cm | charcoal | 1230 ± 25 | GrN-9060 | Versteeg, 1980a |
| | pit 1 125 cm | charcoal | 1210 ± 25 | GrN-9059 | Versteeg, 1980a |
| | pit 1 75 cm | charcoal | 1145 ± 30 | GrN-9360 | Versteeg, 1980a |
| | pit 3 290-295 cm | wood | 1310 ± 30 | GrN-10301 | Unpublished |
| | pit 3 155 cm | charcoal | 1240 ± 30 | GrN-9359 | Versteeg, 1980a |

belonging to the colonial period have been found at the Hertenrits, nor at Prins Bernhard Polder. Calibration of the date of the wooden artefact from Prins Bernhard Polder suggests a presence of Late Hertenrits people in the coastal plain of western Suriname in the 13th century A.D. The possibility cannot be excluded that habitation of the sites associated with this pottery style lasted until the contact period; in that case probably the disturbance associated with these events caused the abandonment of these coastal sites.

## The Buckleburg-1 Mound

Buckleburg-1 is situated ca. 10 km east of the Hertenrits mound (Figure 13:1). Its stratigraphy proved to consist of three main units which are illustrated in figure 13:2. The deepest unit consists of blue/grey clay deposited under marine conditions (Brinkman and Pons 1968; Roeleveld and Van Loon 1979) as indicated by high percentages of *Avicennia* spp. and *Rhizophora* spp. pollen. The upper limits of this unit have a humic character and it is situated about 200 cm under the present surface of the mound. On top of this unit several layers of clay deposited under predominantly freshwater conditons occur. This has been designated unit 2. Its freshwater origin is indicated by the low percentages of *Avicennia* spp. and *Rhizophora* spp. and high percentages of Cyperaceae and *Typha* spp. pollen. This layer is approximately 25 cm thick (Figure 13:2). No artefacts were found below it and the very regular contacts and the structure of the layers indicate that they both consist of naturally deposited clay. Unit 3, occurring at 175/180 cm below the present surface of the mound, consists of grey clay layers with concentrations of artefacts, often characterised by darker coloured clay. These darker coloured areas may have been settlement sites during the building of the mound. Higher percentages of *Avicennia* spp. pollen are recorded in unit 3, indicating that the clay used for constructing the mound was taken predominantly from an environment rich in *Avicennia* pollen.

The ceramics found in the Buckleburg-1 mound proved to be of a different style than pottery from the Hertenrits and Wageningen-1 mounds. Vessel forms and decoration patterns are comparable to the Mabaruma style, earlier found more to the west in north-west Guyana by Evans and Meggers (1960). The Buckleburg pottery (Figures 13:3 and 13:4) is very similar to the "Mabaruma incised" type ascribed to the earliest phase of Mabaruma found in Guyana by Evans and Meggers (1960). Lathrap (1964; 1966) doubted the initial dating of A.D. 500, proposed by Evans and Meggers, suggesting a much earlier dating for this style.

The present study yields direct evidence for the initial dating of Mabaruma ceramics in the Guyanas. Two C14 dates from unit 2 yielded dates of 1835 B.P. and 1735 B.P. (GrN-9804 and GrN-10304). The samples are

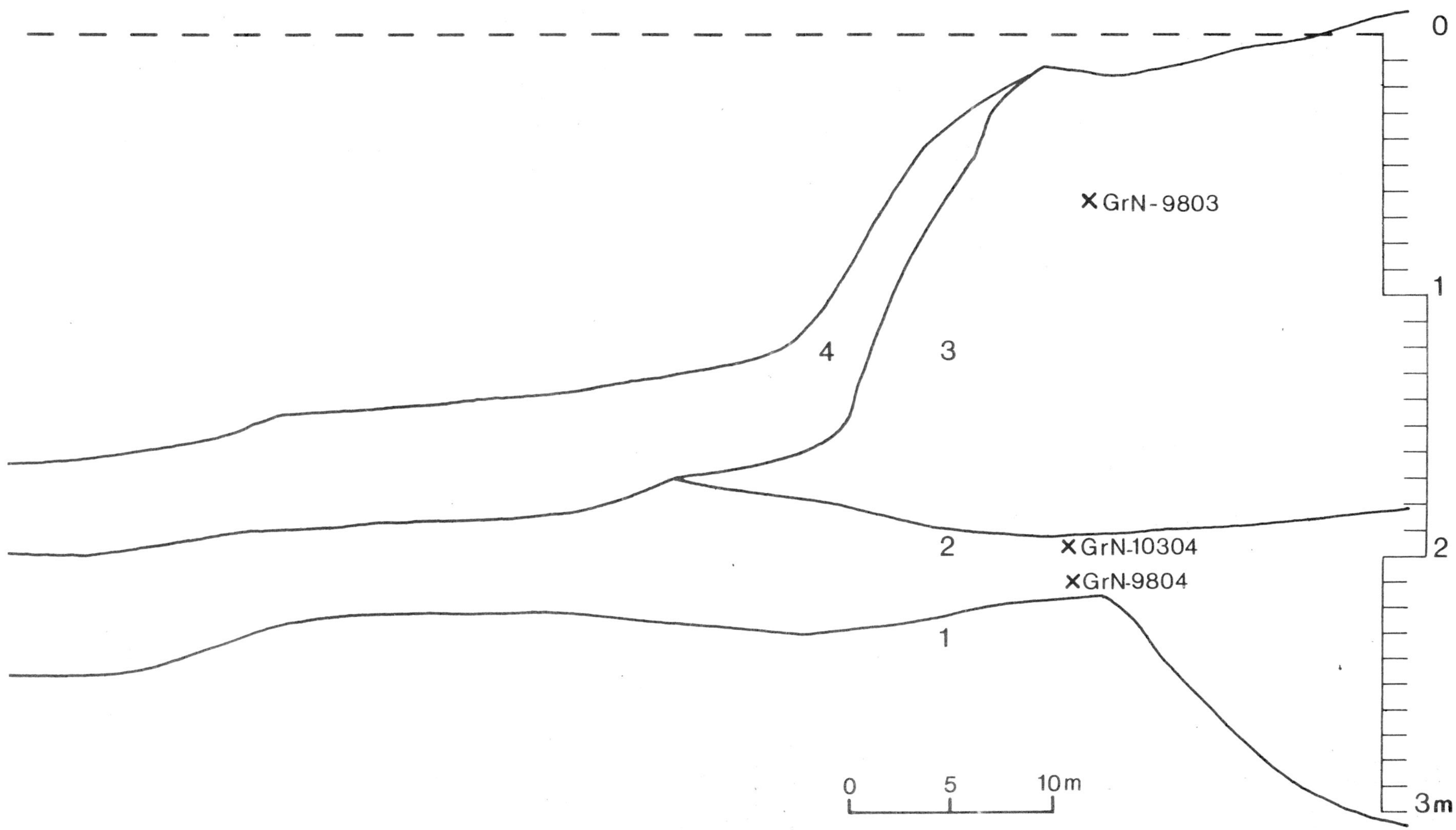


Figure 13:2 Stratigraphical units 1-4 distinguished at Buckleburg-1 x $^{14}C$ samples

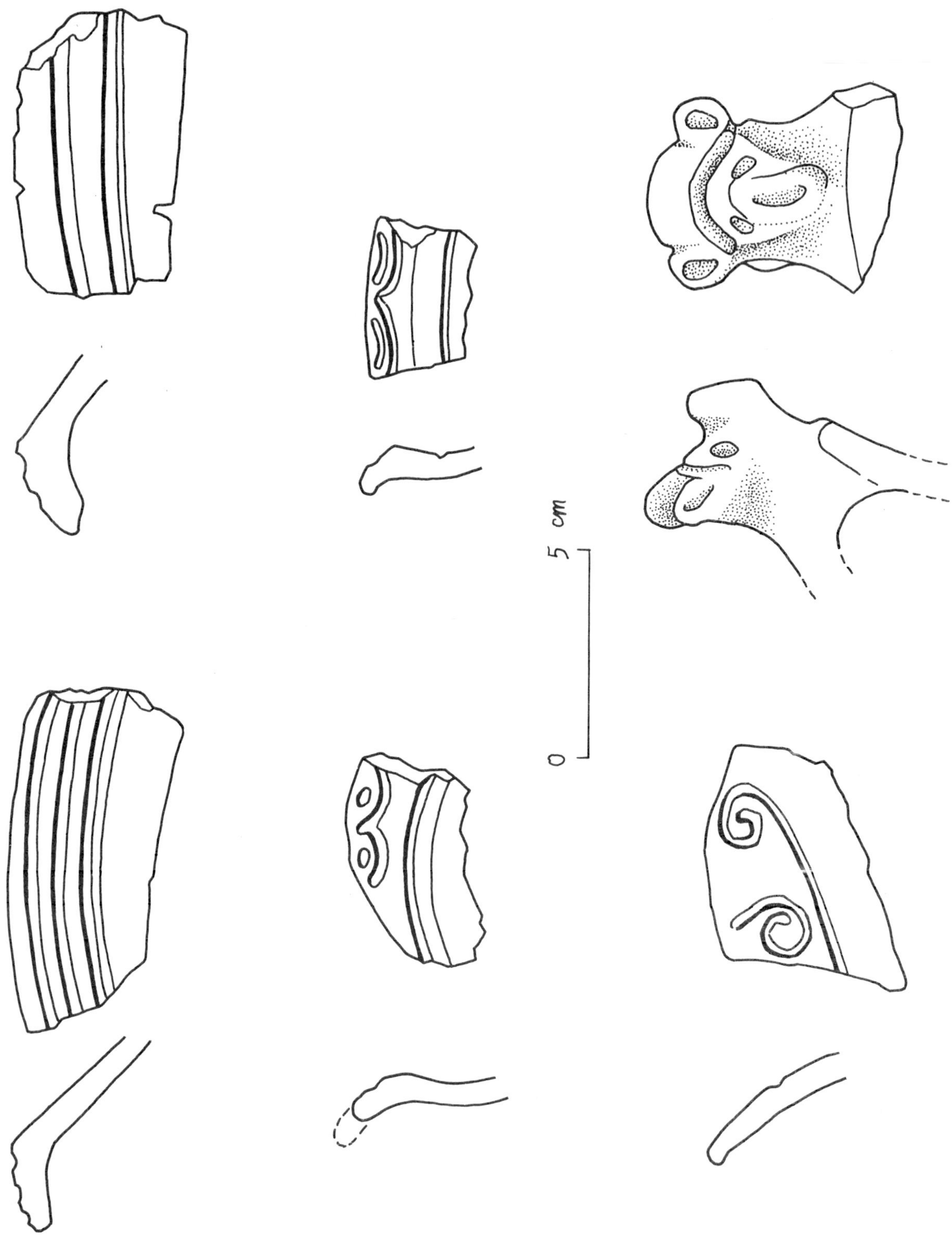


Figure 13:3 Decorated pottery found at Buckleburg-1

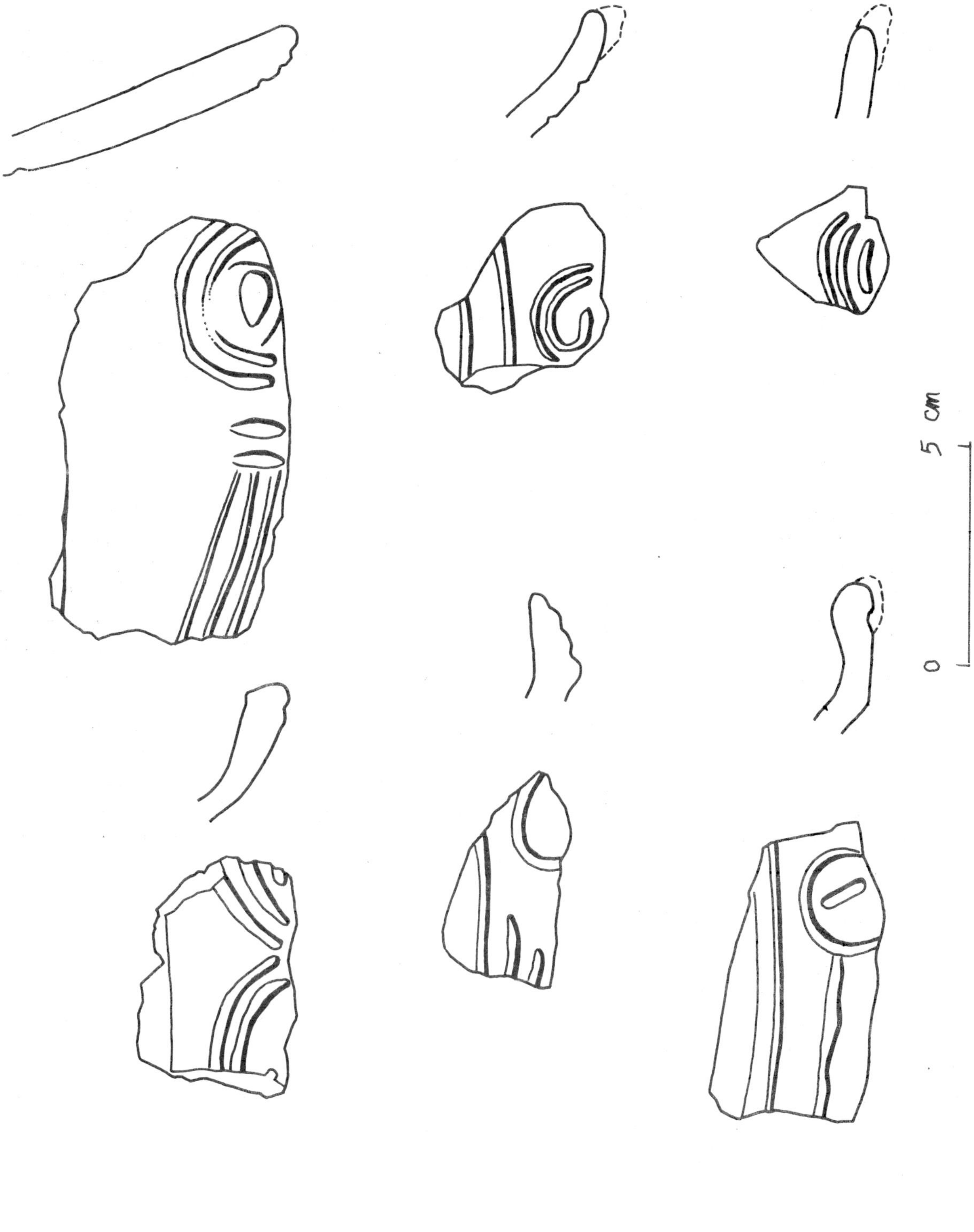


Figure 13:4 Decorated pottery found at Buckleburg-1

stratigraphically and spatially related (Table 13:1) and in view of this these dates are being taken as representative of the first habitation phase at Buckleburg-1.

The pottery found at Buckleburg-1 is similar to pottery found at several Mabaruma sites in Guyana. In the opinion of Evans and Meggers (1960) and Lathrap (1966) the latter represent the earliest phase of Mabaruma in Guyana, so it is likely that the Buckleburg dates are similar to the Mabaruma sites in Guyana. Using the recent calibration list of Klein et al. (1982) with a confidence limit of 95% the first occupation of the mounds occurred between A.D. 15-20 and A.D. 250-255. In addition a $C^{14}$ date obtained from a sample from unit 3 taken from just beneath the contact with unit 4 produced a date of ca. 1300 B.P. (GrN-9803).

Although vessel form and decoration patterns indicate a strong relationship between the pottery found at Buckleburg-1 and that at the Guyana sites, differences in the temper of the pottery are evident. A majority of the Buckleburg pottery has been tempered with crushed pottery while such a temper is not found in the Guyana pottery. Evans and Meggers (1960) have found Mabaruma pottery at sites on hills and river banks in Guyana, while Buckleburg-1 is a mound with associated raised fields in a coastal location. Although mounds (and probably also raised fields) are known to exist in Guyana (Poonai 1962; Goodland 1964; Boomert 1978; Versteeg 1980) particularly between the Canje and Corantijn Rivers, excavations have not been carried out there yet, so it is not known whether they contain Mabaruma ceramics too.

## Conclusions on the Mounds

While previous studies of Suriname's wetlands concentrated on the Hertenrits mounds, recent investigations have yielded new data on two other mounds. At Buckleburg-1 the existence of a mound associated with a different ceramic style and yielding significantly different dates to the Hertenrits has been documented. An Amerindian community who made Mabaruma style ceramics is dated to have existed between ca. 1850 B.P. and 1300 B.P.

Amerindians who produced pottery of the Early Hertenrits style settled at the Wageningen-1 and Hertenrits sites from 1300 B.P. A causal relationship may exist between the end of habitation at Buckleburg and the arrival of new settlers at Wageningen-1 and Hertenrits. The currently available data, however, do not exclude the possibility of a coexistence.

The stratigraphy of the three mounds investigated show strong similarities with three main stratigraphical units being common to each of the sites. A fourth stratigraphical unit occurs in the swamps and on the ramps of the mounds. The matrix of the fourth unit consists of relatively recently deposited sediments of the marine Comowine member.

In view of the parity in the stratigraphy of the mounds, the differences in their datings are striking. All three mounds yielded evidence of a freshwater phase of sedimentation following the end of deposition of the marine Moleson member. At Wageningen-1 and Hertenrits no indications of a long dry period are seen.

The oldest artefacts at Buckleburg-1 have been found in the freshwater dominated unit. Stratigraphy and find patterns indicate that relatively soon after the area became freshwater-dominated, around about 1800 B.P., man arrived and settled somewhere near the location of Buckleburg-1.

The same pattern of events occurred at Wageningen-1 and Hertenrits in ca. 1300 B.P. These events suggest that possibly an accretion of the coast in a north-westerly direction occurred between 1800 and 1300 B.P., between the Buckleburg-1 and Wageningen-1 sites.

## The Raised Fields

The raised fields of the Suriname coastal plain were studied by Boomert (1976). Most of the data were derived from air photographs. Laeyendecker-Roosenburg (1966) discussed a pollen diagram from the raised fields near the Hertenrits. Eight clusters of raised fields have been discovered in the coastal plain of western Suriname. Six clusters are associated with mounds, while two are situated near the Peruvia chenier site (Figure 13:1).

Boomert (1976) distinguishes two types of raised fields in Suriname: (i) square to rectangular plots in geometric patterns with dimensions of the square plots of ca. 4m x 4m, while the rectangular ones are 4m wide and tens of metres long. These fields are constructed in regular patterns. This type has been found near the mound Buckleburg-1 and to the east of the site: (ii) rectangular plots ranging in length from 8 to 140m and in width from 3 to 6.5m are clustered into irregularly arranged groups. A small and a large cluster of this second type has been found near the Hertenrits and Wageningen-1 mounds. The fields are arbitrarily dispersed or are situated close together in irregular groups of two to ten plots (Boomert 1976). One cluster is situated on a low chenier between the Wageningen-1 and Hertenrits mounds. On the basis of their geographical position the arbitrarily dispersed fields are associated with the mounds that yield Hertenrits style pottery. The regularly arranged fields are situated near the older mound Buckleburg-1, which yields Mabaruma style pottery.

All of the raised fields in the cluster in the east of Suriname all belong to the first group. In fact all of

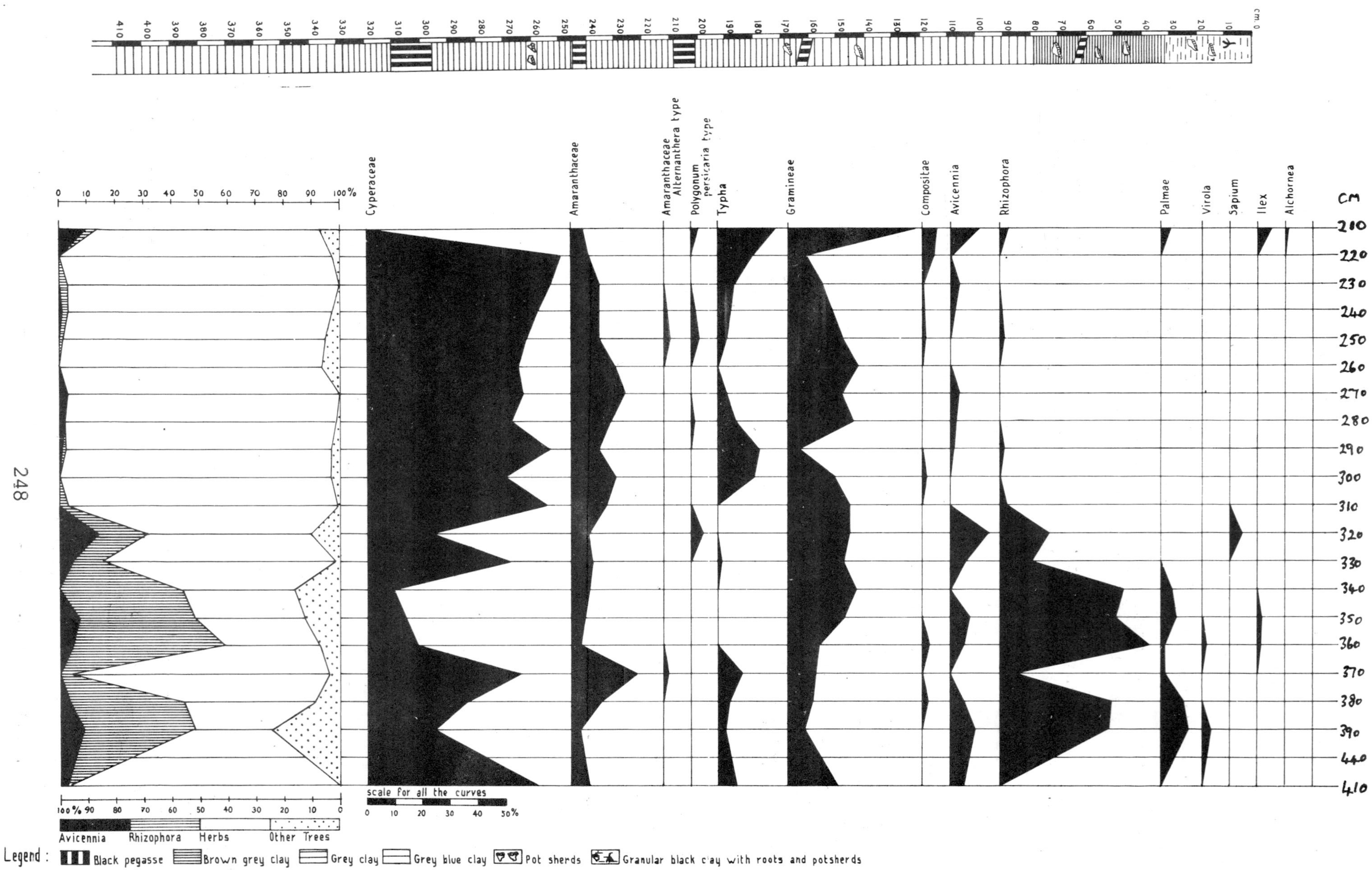


Figure 13:5 Pollen diagram H[3] from the Hertenrits mound. Modified after Laeyendecker-Roosenburg, 1966

the raised fields associated with settlements on chenier ridges belong to this regularly arranged group. Pottery of Hertenrits style has been found. A $C^{14}$ date from the Peruvia site indicates that the earliest habitation coincides with the initial phase of settlement at the Wageningen-1 and Hertenrits mounds.

## Bibliography


BOOMERT, A. 1976, Pre-Columbian raised fields in coastal Surinam, *Proceedings of the Sixth International Congress for the Study of the Pre-Columbian Cultures of the Lesser Antilles*, Guadeloupe, 1975, 134-145, Gainesville.

BOOMERT, A. 1977, Prehistorie, *In, Encyclopedie van Suriname*, edited by C.F.A. Bruining, J. Voorhoeve & W. Gordijn, Amsterdam/Brussel, 506-515.

BOOMERT, A. 1978, Prehistoric habitation mounds in the Canje River area, *Journal of Archaeology and Anthropology* *1*, (1), 44-52.

BRINKMAN, R. & L.J. PONS, 1968, *A Geo-Pedomorphological Classification and Map of the Holocene Sediments in the Coastal Plain of the Three Guianas*, Soil Survey Papers 4, Wageningen.

EVANS, C. & B.J. MEGGERS 1960, *Archaeological Investigations in British Guiana*, Smithsonian Institution Bureau of American Ethnology, Bulletin 177, Washington D.C.

GEIJSKES, D.C. n.d., *Voorlopig verslag van de opgraving Hertenrits*, Rapport Surinaams Museum, Paramaribo.

GOODLAND, E.A. 1964, The Mound, *Journal of the British Guiana Museum and Zoo* *39*, 9-18.

KLEIN, J., J.C. LERMAN, P.E. DAMON & E.K. RALPH 1982, Calibration of Radiocarbon Dates, *Radiocarbon* *24*, (2), 103-150.

LAEYENDECKER-ROOSENBURG, D.M. 1966, A palynological investigation of some archaeologically interesting sections in northwestern Surinam, *Leidse Geologische Mededelingen* *38*, 31-36.

LATHRAP, D.W. 1964, An alternative seriation of the Mabaruma Phase, Northwestern British Guiana, *American Antiquity* *29*, 353-359.

LATHRAP, D.W. 1966, The Mabaruma Phase: a return to the more probable interpretation, *American Antiquity* *31*, 558-566.

POONAI, N.O. 1962, Archaeological sites on the Corentyne coast, Journal of the British Guiana Museum and Zoo 33, 52-53.

ROELEVELD, W. 7 A.J. VAN LOON, 1979, The Holocene development of the Young Coastal Plain of Surinam, Geologie & Mijnbouw 48, 215-224.

VAN DER HEIDE, J., 1973, De Hertenrits, Rapport Dienst Bodemkartering, Paramaribo.

VERSTEEG, A.H. 1980, Prehistoric Cultural Ecology of the Coastal Plain of Western Suriname, Proceedings of the Eighth International Congress for the Study of the Pre-Columbian Cultures of the Lesser Antilles, St. Kitts, 1979, Anthropological Research Papers, 22, 89-98, Tempe.

VERSTEEG, A.H. 1980a, C-14 datings from archaeological sites in Suriname, Mededelingen Surinaamse Musea 32, 38-56.

VERSTEEG, A.H. 1983, Recent archaeological investigations in Suriname, Suralco Magazine 15, (1), 1-9.

VOGEL, J.C. & H.T. WATERBOLK, 1964, Groningen Radiocarbon Dates V, Radiocarbon 6, 349-369.

Chapter 14

# DRAINED FIELD AGRICULTURE AND SOIL FERTILITY

Richard T. Smith

## Abstract

Although much of the present global reserve of arable land is in the tropics some of it was used for intensive agriculture in the past. In particular, large tracts of wet highland and lowland areas in the New World have been used for drained field agriculture. At first sight the swamps and floodplains used are suggestive of problematical environments where agriculture would be limited by alternating conditions of inundation and drought which together induce many pedological problems. Among the soil limitations of such environments are rates and mechanisms of organic matter decomposition, toxicity and nutrient availability, the nitrogen economy, and soil salinity. These factors are discussed and it is shown that in a well managed drained field system they can be overcome or ameliorated. It is suggested that the reason that drained fields have not been found in apparently suitable wetlands is that indigenous peoples could not, for some reason, overcome the soil problems. If archaeologists considered the pedological environments of wetlands the search for drained fields would be aided.

## Resumen

Aunque gran parte de la reserva global actual de suelos cultivables se encuentra en los trópicos, parte de ellos fueron trabajados intensivamente en otra época. Específicamente, grandes trechos de areas de tierras altas y de tierras bajas en el Nuevo Mundo se usaron para la agricultura de campo drenado. A primera vista los pantanos y los llanos de inundación sugieren la existencia de medios ambientes problemáticos en que la agricultura se encontraría limitada por condiciones alternantes de inundación y de sequía, las cuales se combinan para inducir gran número de problemas de suelos. Entre las limitaciones de los suelos de tales medios ambientes figuran las tasas y mecanismos de descomposición de materia orgánica, la toxicidad y disponibilidad de nutrientes, la economía de nitrógeno, y la salinidad de los suelos. Se discuten estos factores y se demuestra que dentro de un sistema de campo drenado bien manejado es posible superarlos o mejorarlos. Se sugiere que la explicación de la ausencia de campos drenados en suelos húmedos que son aparentemente apropiados es que los indígenas no pudieron, por algún motivo, solucionar los problemas de los suelos. La búsqueda de campos drenados se facilitaría si los arqueólogos tomaran en cuenta las características de los suelos de las tierras húmedas.

## Introduction

Most of the unused global potential to increase yields and extend arable land is in tropical Africa and South America (Foth 1982) and yet many tropical regions may once have been extremely productive judging from the remains of long-abandoned intensive agricultural systems such as drained fields and terrace systems documented in recent years. These remains also have strong implications for the technical capabilities and organisation of indigenous cultures whose knowledge of hydrological engineering was vast (Park 1983; Farrington Chapter 12). In the context of drained field agriculture the present known distributions of particular palaeoagricultural forms reflects our incomplete knowledge, arising from the limits of existing surveys and the inevitable obliteration of micro-landforms in the course of time. On the other hand, although an approximate time scale has been suggested for the Prehispanic use of drained fields we must try to distinguish between a perennial or seasonal agricultural role and demonstrate from a knowledge of soil science that such fields would have been able to sustain production over many years at a stretch or even indefinitely.

Drained fields are evidence of attempts to control a range of soil moisture conditions associated with wetlands that are inundated for varying periods of the year and possibly desiccated for some time too. In essence, they represent attempts to extend the period of the year when crops could be grown in a comparatively fertile environment (Siemens Chapter 3) compared with the surrounding *terra firme*. The vast manual efforts involved in their construction, operation and management also suggests that the wetlands were considered fertile agricultural environments. Nutrients and fresh sediments are deposited each year by floodwater or groundwater rich in nutrients and this adds to the soil fertility. Intensive cultivation and flooding also controls weed growth and the build-up of soil-dwelling pathogens. Therefore, the periodically-waterlogged environment has distinct advantages over terrace and milpa forms of agriculture, prone as the latter is to soil exhaustion and erosion. In lowland environments the function of drained fields was probably dominated by intensifying crop production and water control whereas in upland settings the fields were, in addition, probably used for ameliorating soil microclimate (Knapp and Ryder Chapter 11). Experiments by Knapp and Ryder have shown that slight temperature differences brought about by raising the soil surface may have contributed towards the extension of growing periods or the maintenance of night temperatures above certain thresholds, notably that of frost.

Even though new discoveries of drained field systems are still being reported, there are, it is conceded, vast areas of the New World tropics and subtropics where wetlands, seemingly amenable to the construction of drained fields, have not apparently been exploited, nor, so far, are under consideration as possible sites for contemporary development. This could have been due to the lack of sufficient population pressure to warrant their development or to inappropriate environmental conditions. Table 14:1 summarises the considerations appropriate to the prediction and interpretation of drained fields, the primary criterion, of course, being a local population size sufficiently high to justify the work required and a disciplined social order for its maintenance and extension over a period of time. Beyond this, one has to consider in more detail aspects of the water regime of an area and the potential of its soils. The remainder of this paper looks at the nature of these agricultural environments from the standpoint of soil chemistry and agronomy though a number of the statements to be made are necessarily derived from contemporary situations outside actual drained-field localities.

## Water Regime and Soil Texture

A primary consideration in a drained field system is the duration and mode of waterlogging. When aerobic conditions are replaced by anaerobic, i.e. when air-filled porosity is eliminated, all of the dissolved oxygen is consumed, certain transformations occur within 48 hours, while further changes will take place gradually over a period of weeks. For this reason the length of time which soils or soil horizons are subject to waterlogging is important and the first few days after planting are inevitably the most crucial. Waterlogging may be imposed on an otherwise dry soil by flooding and puddling as in some rice paddies, in which case anaerobic conditions slowly migrate downwards from the surface, the water having originally been fully oxidised. On the other hand, in topographically low areas there will tend to be a permanent zone of saturation above which seasonal fluctuations occur as a result of a general rise of groundwater levels and direct rainfall and infiltration (Figure 14:1). However, in the paddies the surface water must remain oxidised at all times or the rice dies.

While runoff would be directed towards drainage lines or canals there would nevertheless be some tendency for each field to develop its own watertable. This would be dependent on two main factors, soil texture and field width or more specifically the edge-to-area ratio (Figure 14:2). Narrow ridged fields are therefore likely to have higher water tables than square or irregular shaped fields. It could be argued, therefore, that longer, narrower fields would also be more likely to meet the requirement of efficient subsoil drainage as well as possibly being easier to construct if any ridging was involved.

An optimum field width to suit the hydrological conditions of each locality could have been achieved by cutting more and more canals until the desired water level

Table 14:1 MAJOR CONTROLS ON DRAINED FIELD DISTRIBUTION

| | General Considerations | Specific Constraints |
|---|---|---|
| 1. | Human history and prehistory | Location in relation to coeval settlement<br>Subsequent destruction |
| 2. | Water regime | Mode and duration of waterlogging<br>Amplitude of seasonal change in water levels<br>Water quality |
| 3. | Physical and chemical nature of substrate | Soil texture and soil morphology<br>Content of organic matter, salts and iron<br>Environmental pH and Eh |

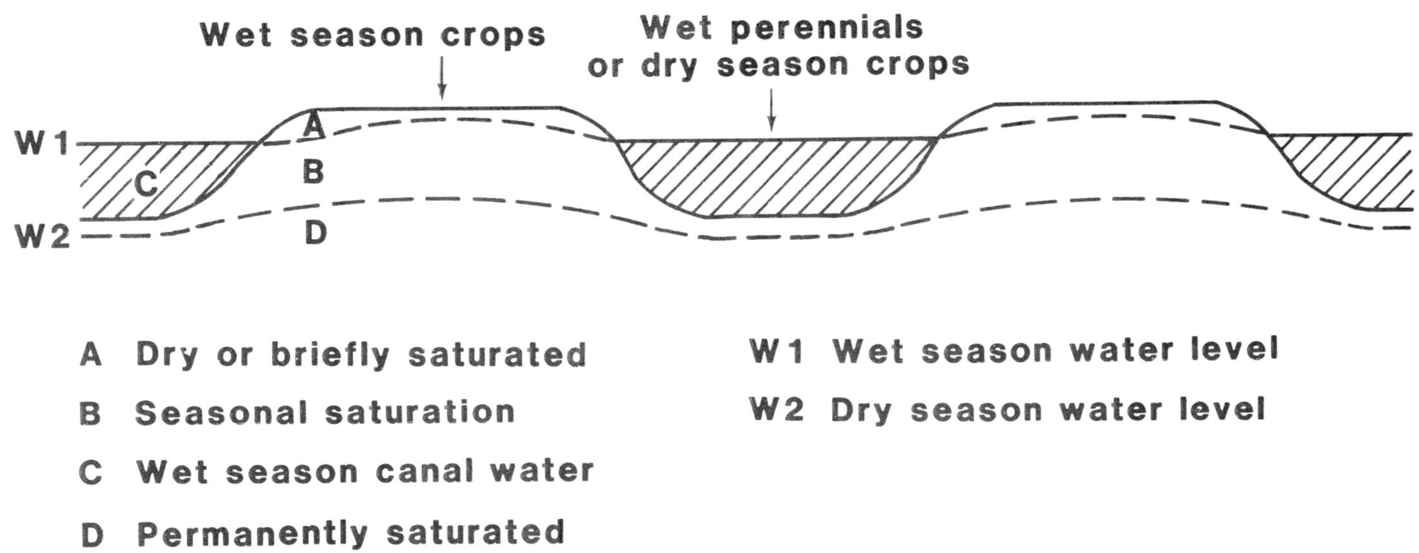


Figure 14:1 Idealised model of seasonal water level change

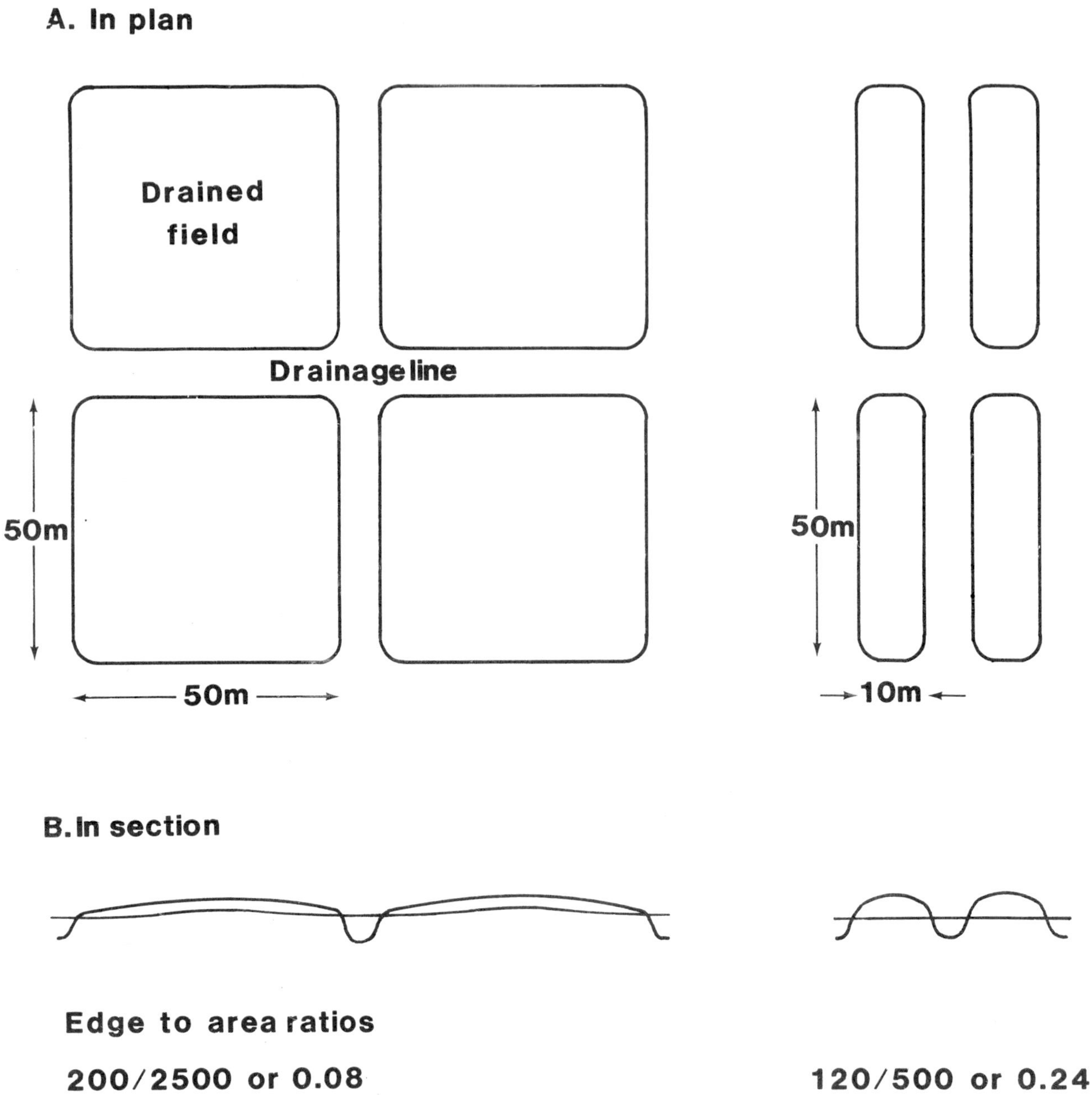


Figure 14:2 Hypothetical relationships between different widths of field and the watertable

was reached, the intensity of drainage may also have depended on how long a period in the year particular areas were required to remain productive.

Experience with particular soils and simple drainage lines appears to have been paralleled by increasing expertise in the operation of canal systems with dams or sluice gates to allow more sophisticated forms of water control. The two canals linking the New River to Pulltrouser Swamp, in Belize, illustrate such a system (Turner and Harrison 1981). Such water control demands some central organisation of the farmers and suggests a powerful governing body, in contrast to which, practices such as maintaining a cambered or convex field cross-section (Wilken 1969) to facilitate runoff, could be achieved on an individual field basis. It is also possible that with control of canal networks many areas might have been effectively irrigated in the dry season by impounding water in canals. Related to this latter point, the clays in many tropical lowland areas, especially those of the Maya area, are of the shrinking-swelling variety. Once these soils dry they become markedly structured and very hard and this leads to difficulties in producing a good tilth (Debano 1969). Therefore, one further reason why wetlands were chosen for arable farming may have been that the clays could be kept permanently moist, fully expanded, and, therefore, more easily workable.

## Decomposition, Toxicity and Nutrient Availability

A soil's capacity to support plant growth and to promote biological processes depends very largely on its chemistry and its water relationships throughout the year. Chemistry is very important. A neutral pH is optimum for biological activity. In base-rich environments supplies of available soil oxygen will, therefore, tend to be used up quickly depending on the rate of diffusion from the surface. Demand for soil oxygen is ultimately dependent on the quantity of organic matter undergoing decay, including crop residues, together with the vigour of root development associated with the present crops or vegetation and soil temperature, which to a large extent controls the rate of decomposition. If all dissolved oxygen is used up on account of waterlogging, decomposition is slowed rather than prevented (Bryant 1971). In such a case aerobic respiration gives way to anaerobic respiration, to which conditions, specific groups of bacteria are adapted. Obligate anaerobes utilise either combined oxygen or other oxidising agents (electron acceptors) for metabolising organic matter. They utilise nitrate ($NO_3^-$), manganese IV ($Mn^{4+}$), iron III ($Fe^{3+}$) and sulphate ($SO_4^{2-}$). One of the first consequences of anaerobic respiration is thus a depletion of soil nitrogen which is lost in gaseous form. If water-logging is maintained, the other electron donors will be progressively consumed and the redox potential will be lowered to the point at which sulphate is reduced to sulphide (Armstrong 1982; Rowell 1981). In a swamp environment this will yield toxic hydrogen sulphide. While most soil nitrogen derives ultimately from the atmosphere via organic matter, sulphur originates from the breakdown of proteins and from the weathering of sulphide minerals. Along coasts and estuaries it is also derived from the ingress of sea water or from seepages from former marine deposits.

These considerations present a gloomy image for the environments characterised by most drained fields yet it is the very proximity to nutrient rich saturated subsoil which confers chemical fertility on drained field soils. It will also determine the potentiality for dry season cropping associated with the topographically low areas. In the first place, accelerated weathering of the saturated zone associated with its lack of oxygen and hence low Eh (Brinkman 1970) and enhanced mobilisation of a number of nutrient ions, including phosphorus, leads to conditions of greater nutrient availability in the soil as a whole. In particular, any roots which are able to follow a falling water table would especially benefit from the latter process for they would be able to obtain iron and phosphate which, as the soil dries out, become much less accessible to uptake. Furthermore, in aerobic respiration there is no general drift of pH. In the case of anaerobic respiration the pH will rise as Eh falls. This may not actually be noticeable in soils of higher pH and the tendency in the case of alkaline soils may even be the reverse owing to a larger production of $CO_2$ (Rowell 1981). This pH effect, though minor, can be sufficient to render more available certain nutrient ions which become fixed at higher pH. However, flooding tends to lower high soil pH values and raise low pH values to about pH 6, a favourable environment for most crops. Again, it is certainly true that the ingress of nutrient-rich floodwater would, by diffusion, even if not direct sedimentation, replenish mineral nutrients consumed by crops as is the case in many rice-growing environments. It might also be argued that this would be more effective the smaller or narrower the fields.

Although cropping takes place principally in aerobic surface soil it will be appreciated that it is impossible to divorce this surface environment from the one beneath and to each side of it when attempting to explain why the drained field system could be viable for long periods. It must also be stressed that a fall in groundwater levels in the dry season is crucial for prevention of soil toxicities in the same way that periodic drying is essential for the somewhat analogous rice paddy soil (De Datta 1981; Patrick and Mahapatra 1968). This periodic oxidation militates against, although it may not totally stop, the production of hydrogen sulphide and hydrocarbon gases, especially inhibitory to root development. In addition it prevents iron and manganese reaching excess levels and becoming toxic for plant growth (Ayotade 1977; Ponnamperuma 1972). The drying of soils also assists in the formation of structural peds which, in a soil profile whose origin may be largely artificial, is important for good aeration and drainage.

The wetland environment might be considered as being problematical but in reality, in a well-maintained drained field system, the seasonal rhythm of wetting and drying can be seen to provide a credible chemical and ecological basis for agriculture.

The Nitrogen Economy

Aside from the more obvious roles of networks of canals and ditches for drainage, irrigation and transportation it is very likely that these were a principle source of soil nitrogen. It is likely that this arose through the clearance and dredging of canals and the application of this highly nitrogenous organic material to the surfaces of drained field plots. The main source of the nitrogen was probably algal slime but other organic debris, subject to its state of decomposition and the timing of its application to field surfaces, would have had a beneficial effect (Wilken 1981). 'Blue-greens' or Cyanophytes, are autotrophic organisms which not only fix carbon but utilise atmospheric nitrogen, including that dissolved in water. An algal associate of the water fern (Anabaena azolla) can thus fix up to 2.5 kg N/ha/day under optimal conditions (Vasey 1981). Blue green algae grow most vigorously when the pH is above neutrality, between 7.5-9.0. Coastal areas and the lowland karst bajos of Yucatan and elsewhere are likely to have provided a suitably alkaline environment even though in a surface environment the pH of an aquatic system would not normally exceed 7.5.

One of the agronomic consequences of heavy applications of nitrogenous fertiliser, especially ammonium nitrate, is that the nitrification process generates hydrogen ions which make the surface soil acidic after a time. However, in areas which were naturally calcareous, or at least buffered to some extent against becoming too acidic, there would be no immediate or long term problem associated with the above practice. Indeed some reduction in alkalinity of certain soils, e.g. vertisols would have been beneficial, for, while having a negligible effect on soil structure any fall in surface pH would have aided the availability of many nutrient elements, as with periodic anaerobism.

It is also likely that leguminous crops grown in rotation, between rows of other crops, or as 'indigenous mixtures' with corn (Gliessman 1979) provided a further source of nitrogen. But the important realisation here is that the apparent disadvantages of the semi-swamp locality for nitrogen balance appear to be more than offset by freely available nitrogen sources.

## Salinity Considerations

The content of salts which occur in groundwater or floodwater, particularly chlorides and sulphates, may locally be of significance and in view of the proximity of a capillary fringe to cultivation surfaces, conditions of high salinity could conceivably be encountered in drained field areas during the dry season. In areas subject to coastal salt influence, fluctuating water tables can lead to the most extreme changes in soil conditions. When waterlogged, the pH of estuarine muds is around neutrality and if this substrate dries or is drained, various saline zones may be produced due to capillarity which depend on the solubility products of the respective ion pairs. Chlorides will often reach the surface as an efflorescence; sulphates, carbonates and iron salts are generally precipitated within soil horizons, each of these salts moving to a lesser degree (Duchaufour 1982).

High salt concentration will increase the osmotic potential of the soil and induce wilting through failure of the plant to take up water. Sulphides in the original mud are oxidised rapidly to sulphates and in the presence of lime, gypsum crystals will form. Without calcium carbonate, sulphuric acid will be formed and the pH may fall below 3. These processes can occur very rapidly, as illustrated by some attempts to reclaim areas of mangrove swamp (Bloomfield and Coulter 1973). It is therefore possible that attempts to build cultivation platforms near tropical coasts would have met with serious problems, ranging from those of salinity to extreme acidity which is equally inimical to plant growth.

Nevertheless, raised field complexes near the coast are reported. At Cerros on the mouth of the New River in Belize, an advanced raised field system occurs with lined canals (Scarborough 1981). Together with this system macroscopic plant remains and pollen occur, which may suggest that saline incursions were overcome and crop growth was successful, at least for a time. It is possible that the canal linings helped in some way to keep salty water away from the fields. In coastal Suriname, Versteeg (Chapter 13) also reports a complex of raised fields and settlements dating from 1800 BP. The fact that in certain coastal areas these systems appear to have been successful may depend on the shortness of the dry season, in which the effects of soil ripening and capillarity would be minimised. Alternatively a higher lime content than in other tropical areas or the use of lined canals with high embankments may have circumvented the problem.

## Conclusions

This paper has focused on Mesoamerica. With previous experience of forest clearance and farming in areas of relatively higher elevation, exploitation of lowland forests and swamps by the Maya might have appeared reasonably straightforward once forest areas were cleared.

But, the removal of trees leads to a marked rise of water tables so that initial attempts at settlement and the extension of agriculture to areas which we now associate with the Maya lowlands, may well have been checked until methods were available to effectively control inundation. An important over-riding factor would undoubtedly have been the Classic Maya ascendancy, representing a maximum degree of integration not only of communications but of drainage systems as well.

As already discussed, certain factors are to be identified either individually or in combination as favouring or discouraging drained field agriculture. Antagonistic influences might locally include prolonged high ground water levels or marine influence. On the other hand, areas with extremes of water levels between seasons might make for difficult planning of appropriate hydraulic structures. High clay content, high salinity and low iron contents are likely to present problems. Particularly favourable influences on the other hand would be provided by sediment deposition following flooding, adequate iron and low salinity. The presence of solid rock or aggregate, especially limestone, would provide ballast and therefore aid in field building and preparation; abundant organic matter applications to the field surfaces, would generally be a beneficial fertilizer provided drainage is maintained. The main issue is, however, considered to be the maintenance of a suitable soil aeration state.

Archaeologists are familiar with the problem of different kinds of biological or artefactual remains being preserved or destroyed depending on preservation conditions. Similarly, the field archaeologist may be better able to answer such questions as 'should drained fields be found in this area?' if the soil and hydrological conditions of wetlands are understood. Conventional air photography and other forms of remote sensing have recently led to the discovery of many complexes of drained fields which have remained hidden for centuries. These data are allowing the update of relict distribution maps of drained fields, yet they are incomplete records of the past. Scientific principles need to be applied not only to enlarge our understanding of how former agricultural systems functioned but to determine ecological factors which would have limited their true geographical range.

## Bibliography

ARMSTRONG, W. 1982, Waterlogged soils, In, Environment and Plant Ecology, edited by J.R. Etherington, John Wiley & Sons, Chichester, Ch.10. 290-330.

AYOTADE, K.A. 1977, Kinetics and reactions of hydrogen sulphide in solutions from flooded rice soils, Plant and Soils 46, 381-389.

BAVER, L.D., W.H. GARDNER and W.R. GARDNER 1972, *Soil Physics*, Wiley, New York, 498pp.

BLOOMFIELD, C. and J.K. COULTER, 1973, Genesis and management of acid sulphate soils, *Advances in Agronomy* 25, 265-326.

BRINKMAN, R. 1970, Ferrolysis: a hydromorphic soil forming process, *Geoderma* 3, 199-206.

BRYANT, C. 1971, *The Biology of Respiration*, Institute of Biology Studies in Biologu No.28, Edward Arnold, London, 60pp.

DEBANO, L.F. 1969, Water-repellant soils. *Agricultural Science Review* 7, (2), 11-18.

DE DATTA, S.K. 1981, *Principles and Practices of Rice Production*, John Wiley and Sons, Chichester, 380pp.

DUCHAUFOUR, P. 1982, *Pedology: Pedogenesis and Classification*, Allen and Unwin, London, 448pp.

FOTH, H.D. 1982, Soil resources and food: a global view, *In, Principles and Applications of Soil Geography*, edited by E.M. Bridges and D.A. Davidson, Longman, London, Ch.8, 256-274.

GLIESSMAN, S.R. 1979, Some ecological relationships of traditional agro-ecosystems in the lowland humid tropics of Southern Mexico - Mimeo, Paper presented at 43rd International Congress Americanists, Vancouver.

HILLEL, D. 1982, *Introduction to Soil Physics*, Academic Press, New York, 364pp.

PARK, C.C. 1983, Water resources and irrigation in pre-Hispanic Peru, *Geographical Journal*, 149, (2), 153-166.

PATRICK, W.H. and MAHAPATRA, I.C. 1968, Transformation and availability to rice of nitrogen and phosphorus in waterlogged soils, *Advances in Argronomy* 20, 323-358.

PONNAMPERUMA, F.N. 1972, The chemistry of submerged soils, *Advances in Agronomy* 24, 29-96.

ROWELL, D.L. 1981, Oxidation and reduction, *In, The Chemistry of Soil Processes*, edited by D.J. Greenland and M.H.B. Hayes, John Wiley and Sons, Chichester, 401-461.

SCARBOROUGH, V.L. 1981, A preclassic Maya water system, Paper presented to American Society Archaeologists, Minneapolis.

TURNER, B.L. II and P.D. HARRISON 1981, Prehistoric raised-field agriculture in the Maya lowlands, Science 213, 399-405.

VASEY, D.E. 1981, Nitrogen fixation and flow in experimental island bed gardens: implications for archaeology, Paper presented at the conference on Prehistoric intensive agriculture in the tropics, Canberra, To be published B.A.R. Oxford 1983.

WILKEN, G.C. 1969, Drained-field agriculture: an intensive farming system in Tlaxcala, Mexico, Geographical Review 59, (2), 215-241.

WILKEN, G.C. 1981, A note on buoyancy and other dubious characteristics of the 'floating' chinampas of Mexico, Paper presented at the conference on Prehistoric intensive agriculture in the tropics, Canberra, To be published B.A.R. Oxford, 1983.